Lecture Notes in Computer Science 9951

Commenced Publication in 1973
Founding and Former Series Editors:
Gerhard Goos, Juris Hartmanis, and Jan van Leeuwen

More information about this series at http://www.springer.com/series/7408

Dieter Fiems · Marco Paolieri
Agapios N. Platis (Eds.)

Computer Performance Engineering

13th European Workshop, EPEW 2016
Chios, Greece, October 5–7, 2016
Proceedings

Springer

Editors
Dieter Fiems
Ghent University
Gent
Belgium

Marco Paolieri
University of Florence
Firenze
Italy

Agapios N. Platis
University of the Aegean
Chios
Greece

ISSN 0302-9743 ISSN 1611-3349 (electronic)
Lecture Notes in Computer Science
ISBN 978-3-319-46432-9 ISBN 978-3-319-46433-6 (eBook)
DOI 10.1007/978-3-319-46433-6

Library of Congress Control Number: 2015946767

LNCS Sublibrary: SL2 – Programming and Software Engineering

This Springer imprint is published by Springer Nature
The registered company is Springer International Publishing AG
The registered company address is: Gewerbestrasse 11, 6330 Cham, Switzerland

Preface

It is our pleasure to present the proceedings of EPEW 2016, the 13th European Performance Engineering Workshop, held October 5–7, 2016 in Chios, Greece.

The goal of this annual workshop series is to gather academic and industrial researchers working on all aspects of performance engineering. The papers presented at the workshop reflect the diversity of modern performance engineering, with topics ranging from the analysis of queueing networks and stochastic processes, to performance analysis of computer systems and networks, and even modeling of human behavior.

The call for papers gathered 25 submissions by authors from 13 countries. Each paper was peer reviewed by an average of three reviewers from the Program Committee (PC) on the basis of its relevance, novelty, and technical quality. After the collection of reviews, PC members discussed the quality of the submissions for one week before deciding to accept 14 papers.

This year, we were honored to have two keynote speakers. Prof. Kishor S. Trivedi from Duke University (USA) addressed current research on the quantitative analysis of network survivability. Prof. Nicholas Ampazis from the University of the Aegean (Greece) explored the use of deep learning approaches for performance analysis.

We thank our keynote speakers, as well as all PC members and external reviewers, who returned their reviews on time despite the tight reviewing deadline, and provided constructive and insightful comments. We also express our gratitude to the Organizing Committee at the University of the Aegean for their continuous and valuable help, the EasyChair team for their conference system, and Springer for their continued editorial support. Above all, we would like to thank the authors of the papers for their contribution to this volume, which we hope that you, the reader, will find useful and inspiring.

August 2016

Dieter Fiems
Marco Paolieri
Agapios N. Platis

Organization

Program Committee

Simonetta Balsamo	University of Venice Ca' Foscari, Italy
Marta Beltran	King Juan Carlos University, Spain
Marco Bernardo	University of Urbino, Italy
Laura Carnevali	University of Florence, Italy
Giuliano Casale	Imperial College London, UK
Vittorio Cortellessa	University of L'Aquila, Italy
Tadeusz Czachorski	Polish Academy of Sciences, Poland
Dieter Fiems	Ghent University, Belgium
Jean-Michel Fourneau	University of Versailles, France
Stephen Gilmore	University of Edinburgh, UK
András Horváth	University of Turin, Italy
Gábor Horváth	Budapest University of Technology and Economics, Hungary
Alain Jean-Marie	CNRS University of Montpellier, France
Helen Karatza	Aristotle University of Thessaloniki, Greece
William Knottenbelt	Imperial College London, UK
Spyros Kokolakis	University of the Aegean, Greece
Samuel Kounev	University of Würzburg, Germany
Vasilis Koutras	University of the Aegean, Greece
Lasse Leskelä	Aalto University, Finland
Catalina M. Lladó	University of the Balearic Islands, Spain
Andrea Marin	University of Venice Ca' Foscari, Italy
Marco Paolieri	University of Florence, Italy
Roberto Pietrantuono	University of Naples Federico II, Italy
Agapios Platis	University of the Aegean, Greece
Philipp Reinecke	HP Labs Bristol, UK
Anne Remke	WWU Münster, Germany
Sabina Rossi	University of Venice Ca' Foscari, Italy
Markus Siegle	Bundeswehr University Munich, Germany
Miklos Telek	Budapest University of Technology and Economics, Hungary
Nigel Thomas	Newcastle University, UK
Petr Tuma	Charles University, Czech Republic
Maaike Verloop	CNRS University of Toulouse, France
Jean-Marc Vincent	Joseph Fourier University, France
Demosthenes Vouyioukas	University of the Aegean, Greece

Joris Walraevens   Ghent University, Belgium
Katinka Wolter   Freie Universität Berlin, Germany
Armin Zimmermann   Technische Universität Ilmenau, Germany

Additional Reviewers

Nikolas Roman Herbst   University of Würzburg, Germany
Jóakim von Kistowski   University of Würzburg, Germany
Jürgen Walter   University of Würzburg, Germany

Invited Talks

Survivability Quantification for Networks

Kishor S. Trivedi

Department of Electrical and Computer Engineering,
Duke University, Durham, NC 27708, USA
ktrivedi@duke.edu

Abstract. Survivability is a critical attribute of modern computer and communication systems. The assessment of survivability is mostly performed in a qualitative manner and thus cannot meet the need for more precise and solid evaluation of service loss or degradation in presence of failure/attack/disaster. This talk addresses the current research status of quantification of survivability. First, we carefully define survivability and contrast it with traditional measures such as reliability, availability and performability [2, 8, 7]. We use "survivability" as defined by the ANSI T1A1.2 committee – that is, the transient performance from the instant an undesirable event occurs until steady state with an acceptable performance level is attained [1]. Thus survivability can be seen as a generalization of recovery after a failure or any undesired event [3]. We then discuss probabilistic models for the quantification of survivability based on our chosen definition. Next, three case studies are presented to illustrate our approach. One case study is about the quantitative evaluation of several survivable architectures for the plain old telephone system (POTS) [5]. The second case study deals with the survivability quantification of communication networks [4] while the third is that of smart grid distribution automation networks [6]. In each case hierarchical models are developed to derive various survivability measures. Numerical results are provided to show how a comprehensive understanding of the system behavior after failure can be achieved through such models.

References

1. ANSI T1A1.2 Working Group on Network Survivability Performance: Technical report on enhanced network survivability performance. ANSI, Tech. Rep. TR No. 68 (2001)
2. Avizienis, A., Laprie, J., Randell, B., Landwehr, C.E.: Basic concepts and taxonomy of dependable and secure computing. IEEE Trans. Dependable Sec. Comput. 1(1), 11–33 (2004)
3. Heegaard, P.E., Helvik, B.E., Trivedi, K.S., Machida, F.: Survivability as a generalization of recovery. In: 11th International Conference on the Design of Reliable Communication Networks. DRCN 2015, pp. 133–140 (2015)
4. Heegaard, P.E., Trivedi, K.S.: Network survivability modeling. Computer Netw. 53(8), 1215–1234 (2009)
5. Liu, Y., Mendiratta, V.B., Trivedi, K.S.: Survivability analysis of telephone access network. In: 15th International Symposium on Software Reliability Engineering. ISSRE 2004, pp. 367–378 (2004)

6. Menasché, D.S., Avritzer, A., Suresh, S., Leão, R.M.M., de Souza e Silva, E., Diniz, M.C., Trivedi, K.S., Happe, L., Koziolek, A.: Assessing survivability of smart grid distribution network designs accounting for multiple failures. Concurrency and Comput. Pract. Exp. **26** (12), 1949–1974 (2014)
7. Meyer, J.F.: On evaluating the performability of degradable computing systems. IEEE Trans. Comput. **29**(8), 720–731 (1980)
8. Trivedi, K.S.: Probability and Statistics with Reliability, Queuing, and Computer Science Applications. Wiley (2001)

Deep Learning Models for Performance Modelling

Nicholas Ampazis

Department of Financial and Management Engineering,
University of the Aegean, Chios 82100, Greece
n.ampazis@fme.aegean.gr

Abstract. Deep learning approaches to performance modelling and prediction of computer systems can be considered as a "black-box" approach, where many layers of information processing stages in hierarchical neural networks architectures are exploited for feature learning in prediction or classification tasks. Examples of deep learning applications to performance modelling span from anomaly detection to optimization, to capacity planning, and, with the advent of cloud computing, to automatic resource provisioning.

Keywords: Deep learning · Machine learning · Neural networks · Performance modelling

1 Introduction

Deep Learning (DL) [1] is a rapidly growing discipline that, during the last few years, has revolutionalised machine learning and artificial intelligence research due to the availability of "big data", new algorithms for neural networks training, and extremely fast dedicated hardware. Companies like Google, Microsoft, Amazon, Facebook and Apple use deep learning to solve difficult problems in areas such as speech and image recognition, machine translation, natural language processing, resource planning or even to reduce power consumption by manipulating computer servers and related equipment like cooling systems [2].

The essence of DL is to compute hierarchical features or representations of observational data, where the higher-level features or factors are defined from primary lower-level measurements. Based on the features extracted from the data in the training set, the calculations within the model are adjusted so that known inputs produce desired outputs. The theory then extends to the fact that, similarly to classical machine learning, a trained deep learning system will correctly recognize the patterns when presented with new examples [7].

Deep learning can be seen as a more complete, hierarchical and a "bottom up" way for feature extraction without human intervention. In the past manually designed features were used in demanding tasks such as, for example, image and video processing. These rely on human domain knowledge and it is hard to manually tune them. Thus, developing effective features for new applications was a slow process. Deep learning overcomes this problem of feature extraction by adaptively determining operator coefficients, like for example in convolutional layers which are exceptionally good at discovering and extracting features from data. These features are propagated to the next layer to form a hierarchy of nonlinear features that grow in complexity (e.g. in an image processing task, from blobs/edges → noses/eyes/cheeks → faces). The final layer uses all these generated features for classification or regression. Deep learning can be thought of as "feature engineering" done automatically by algorithms [3, 6].

2 Applications

In applications of classical Machine Learning (ML) methods to performance modeling or prediction, it was sufficient to identify the core inputs (features) of the performance functions, and the ML algorithm would take care of inferring how they map to target Key Performance Indicators (KPI). Such models are built on the basis of a so called training phase, during which the application is tested with different workloads and is parameterized with different configurations, with the purpose of observing the corresponding achieved performance. Thus their advantage is that the task is reduced to fitting the input data to their desired output values without exploiting any additional knowledge about the application. However input features have to be manually crafted, e.g. small versus large jobs to encode workload intensity, number and types of servers to encode infrastructure, etc. Similarly, KPI outputs like throughput (e.g. max jobs/sec), response time (e.g. execution time of a job) or consumed energy (e.g. Joules/job) would have to carefully defined in order to discriminate the task as being a regression, a classification or a clustering problem.

Relative to other machine learning techniques, DL has four key advantages:

- It can detect complex relationships among features
- It can extract new low-level features from minimally processed raw data
- It can handle multiclass problems with high-cardinality
- It can produce results with unlabeled data

These four strengths suggest that deep learning can produce useful results where other methods may fail. It may also build more accurate models than other methods, and it can reduce the time needed to build a useful model.

Already DL is utilized in order to solve highly practical problems in all aspects of business. For example:

- Payment systems providers use DL to identify suspicious transactions in real time [5].
- Organizations with large data centers and computer networks use DL to mine log files and detect threats [8].
- Vehicle manufacturers and fleet operators use DL to mine sensor data to predict part and vehicle failure [9].
- Deep learning helps companies with large and complex supply chains predict delays and bottlenecks in production [4].

With the increased availability of deep learning software and the skills to use it effectively, we expect the list of commercial applications to grow rapidly in the next several years.

References

1. Bengio, I.G.Y., Courville, A.: Deep Learning. MIT Press (2016, in press)
2. Clark, J.: Google cuts its giant electricity bill with DeepMind-powered AI (2016). http://www. bloomberg.com/news/articles/2016-07-19/google-cuts-its-giant-electricity-bill-with-deepmind-powered-ai. Accessed 19 July 2016
3. Dettmers, T.: Deep learning in a nutshell: core concepts (2015). https://devblogs.nvidia.com/parallelforall/deep-learning-nutshell-core-concepts/. Accessed 18 May 2016
4. Ge, L.: Diving into deep learning: what deep learning could mean for the industrial economy (2015). http://gelookahead.economist.com/deep-learning. Accessed 27 May 2015
5. Harris, D.: How PayPal uses deep learning and detective work to fight fraud (2015). https://gigaom.com/2015/03/06/how-paypal-uses-deep-learning-and-detective-work-to-fight-fraud. Accessed 3 March 2015
6. Jaokar, A.: Evolution of deep learning models (2015). http://www.opengardensblog.futuretext.com/archives/2015/07/evolution-of-deep-learning-models.html?attest=truewpmp_tp=1. Accessed 17 May 2016
7. Kamadolli, S.: Getting real with deep learning (2015). https://medium.com/@kamadoll/getting-real-with-deep-learning-3b7ef698766d#.muip7g5o3. Accessed 18 May 2016
8. Najafabadi, M.M., Villanustre, F., Khoshgoftaar, T.M., Seliya, N., Wald, R., Muharemagic, E.: Deep learning applications and challenges in big data analytics. J. Big Data 2(1), 1–21 (2015)
9. Okanohara, D.: Deep learning in real world: automobile, robotics, bio science (2015). http://hiroshi1.hongo.wide.ad.jp/hiroshi/files/internet/okanohara_2016.pdf. Accessed 27 June 2016

Contents

Modeling and Analysis of Human Behavior

Modeling and Simulation Tools

Analysis and Fitting Methods

Analytic Solution of Fair Share Scheduling in Layered Queueing Networks

Lianhua Li[(⊠)] and Greg Franks

Carleton University, Ottawa, ON, Canada
{Lianhua,greg}@sce.carleton.ca

Abstract. Fair share scheduling has been widely used in many distributed systems. Layered Queueing Networks (LQN) are a widely used performance evaluation technique for distributed systems. Therefore, being able to evaluate performance of systems using fair share scheduling is essential. However, Fair share scheduling in a LQN model could only be solved using simulation previously. A main concern of simulation is long execution times. This paper uses a method called 'Dynamic Parameter substitutions' (DPS) to solve the Fair share scheduling analytically. DPS is an iterative method to calculate state-based parameters using performance results that are found using Mean Value Analysis (MVA). The paper shows how DPS is integrated into the LQNS solver (LQNS-DPS), which makes solutions of models with fair scheduling both fast and scalable. LQNS-DPS was verified using two sets of models, both with cap and guarantee shares. Over 150 randomly parameterized models, throughput found using LQNS-DPS was on average no worse than 6 % of the result found from simulation.

1 Introduction

Fair share (*FS*) scheduling was introduced by Kay and Lauder [13] to prevent allocating resources inappropriately in a student computing center environment. In FS scheduling, a value is assigned to each group of users who have the right to consume a resource. This value specifies how much resource a group is allowed to have; an entitlement of resources is referred to as a share. A FS scheduler distributes the usage of a resource to each task or group based on its share. FS scheduling is intended to prevent tasks or processes from consuming more resources than that they are supposed to, or even to monopolize some resources [27].

Many algorithms to achieve the goal of the fairly sharing resource have been developed. Fair queueing [5,18] and its variants [10,22] are used in all sorts of communication networks, (packets [1,5,18,22] or wireless [21]). Proportional share scheduling [23], includes lottery scheduling [23], stride scheduling [24] (deterministic allocation), Surplus Fair Share (*SFS*) [3], and Completely Fair Share (*CFS*) [20]. Some other algorithms exist, for example, the Borrowed Virtual Time (*BVT*) [6] in virtual machine environments.

© Springer International Publishing AG 2016
D. Fiems et al. (Eds.): EPEW 2016, LNCS 9951, pp. 3–17, 2016.
DOI: 10.1007/978-3-319-46433-6_1

FS scheduling has been supported in some operating systems, for example, a version of the Unix operating system supported it as early as 1984 [11], and Completely Fair Share (CFS) [20] scheduling was adopted into the Linux kernel in the release of 2.6.23. The impact of FS scheduling in virtual machine environments has been studied in [4,7,25,26].

Since the decisions of a fair share scheduler depend on the system's performance at each specific moment, queueing-based performance evaluation techniques have difficulty to analytically evaluate the performance impacts of a fair share scheduler to the system. Layered Queueing Networks (LQN), as extended queueing networks, also face the same difficulty. Accommodating FS scheduling into LQN models was studied in [14], in which CFS [20] was implemented in the PARASOL Simulation Engine [19] and the LQN simulator, LQSIM. The solution in [14] was a simulation solution, so scalability was the main concern. The study of a hybrid solution in the LQN solver [15] was initially motivated to solve FS scheduling in the LQN solver faster than the simulation solution in [14]. In the hybrid solution, simulation was used to solve submodels where processors with FS scheduling reside; all other submodels were solved by the analytic solver LQNS. However, this approach did not work well when the decomposed model had traffic dependencies [16] between the submodels called interlocking.

This paper describes a method called dynamic parameter substitution (DPS) for solving models with FS scheduling analytically using layered queueing. Dynamic parameter substitution (DPS) is an iterative solution that updates the parameters of the underlying queueing networks of a LQN model as the model is being solved. This approach is similar to the method used by analytic solvers that use decomposition to solve models. However, DPS uses functions to adjust values rather than simple substitutions. Therefore, some state-based behaviours can be solved using the performance results that are found by Mean Value Analysis (MVA), without calculating state probabilities. The DPS functions for a FS scheduler are developed based on the ideas of the CFS scheduler used in [14] and listed in Sect. 3.2.

Another challenge for the analytic solution of FS scheduling is the priority issue when a surplus share exists. Groups with a guarantee share may consume more of a resource when a surplus exists. However, granting surplus share to some groups may increase queueing delays of requests from other groups. The reason for this effect is that increasing the share of a group implicitly increases the priority of this group. This work describes a new type of server queue for the LQNS solver, namely 'FairShareServer', to overcome this priority issue.

The rest of this paper is organized as follows: Sect. 2 describes briefly the LQN model. Section 3 describes how fair share scheduling is handled by simulation and the hybrid method in LQN models previously. Next, the method of dynamic parameter substitution and the mappings are explained in Sect. 4. The implementation of the DPS method into the LQNS solver is described in Sect. 5 followed by a comparison of the results of DPS to the results of simulation in Sect. 6. Finally, Sect. 7 summarizes the paper.

2   Layered Queueing Networks

A layered queueing network is a type of extended queueing network that is particularly suitable at solving performance models of modern software systems designed using the client-server paradigm. With this type of software system, clients of servers block awaiting replies to requests. These servers may in turn make requests to other servers. Layering arises by following the call chain from the top-most client all the way down to the lowest level server.

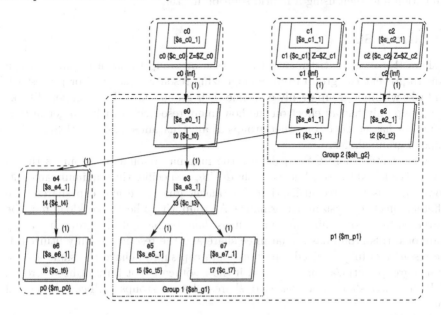

Fig. 1. A randomly generated LQN model.

A layered queueing network, for example Fig. 1, is a directed graph consisting of processors, tasks, entries and requests. *Processors* are shown using rounded rectangles and are used to consume time in the model. The tasks running on the processor are shown inside the rectangle. The number of cores for a processor is shown using a label with braces, for example {m_p1}. *Groups* are used to collect tasks together for the purpose of FS scheduling and are shown using rectangles within a processor. All of the tasks within a group are contained within a rectangle. *Tasks* are shown using large parallelograms and represent autonomous entities. Tasks which accept no requests (for example, c0, c1 and c2) represent clients and are used to generate traffic. Tasks which accept requests represent operating system processes, devices except processors and modeling artifacts. Tasks shown with using a stacked icon represent multi-servers (or in the case of clients, multiple customers). The number of instances is again shown using a label with braces. *Entries* are shown using the parallelograms within tasks and are used to differentiate service. Service times are shown using labels with square brackets, for example [s_c0_1]. Finally, *requests* are made from entry to entry and are shown using the directed arcs. The mean number of requests from one entry to another is shown using a label with parenthesis, for example (1). A more complete description of the model is found in [8].

3 Fair Share Scheduling in LQN

Accommodating FS scheduling into LQN models was first studied in [14] using simulation and then using a hybrid solution in [15].

3.1 Group Shares

A share [13] is the entitlement of each task or group to use a resource, which aims to prevent the group from monopolizing resources (channel or processor). All tasks within a group evenly consume their group share. The number of tasks of a group only has an effect on the how many resources a task can get in that group but has no effect on other groups at all. Guarantee and cap [1,2] are two kinds of group shares.

A guarantee share $GS_{(g,i)}$ defines the minimal amount of resources that a group i is allowed to use while a cap share $GS_{(c,i)}$ specifies the maximum amount. However, it is resource utilization that makes guarantee and cap shares have different effects on system performance. A group with a heavy workload must be prevented from consuming more resources than its share, if this group has a cap share or if this group has a guarantee share and the resource is fully utilized. If the resource is fully utilized, guarantee and cap share behave the same way to limit a group's utilization. However, if the resource is not fully utilized, groups with guarantee shares may get extra shares whereas groups with cap shares will not.

In LQN models, all tasks using a processor with a FS scheduler are grouped into LQN groups [14]. Then, a group share (GS_i) is assigned to each group i; and all the tasks within the group i share or consume CPU time evenly. In LQN models, group shares are specified as a fraction of a resource to which groups are allowed to use; the value of a GS is between 0 and 1 (inclusive). The total share of all groups is one. Therefore, group shares may need to be normalized if the sum of all group shares specified in an input file does not equal to 1 [14]. Guarantee shares and cap shares are scaled in the same rate, shown in (1).

$$GS_i' = \frac{GS_i}{\sum GS_i}, \sum GS_i > 1 \tag{1}$$

where: GS_i' is the group share of the group i after normalization.

3.2 Fair Share Scheduling by LQSIM

Accommodating FS scheduling into LQN models was studied in [14], in which CFS [20] was implemented in the PARASOL simulation engine [19] and the LQN simulator, LQSIM. The principles used for this implementation are:

1. Every group has a priority based on its share and how long its tasks have been waiting.
2. If a group occupies more resources than it is entitled to, all the tasks within the group are suspended (forced to sleep) for an amount of time in order to let other groups consume some of the resources.
3. The amount of sleep time is proportional to the amount of excessive resources it has consumed. Tasks in a suspended group re-enter the ready-to-run-queue of the FS processor after the sleep time of the group is exhausted.

The solution in [14] was a simulation solution, so scalability was the main concern.

4 Dynamic Parameter Substitutions

The principle idea behind dynamic parameter substitution is that some input parameters change as the model is being solved. The key point of handling FS processors is calculating the *extra* delays of the groups after they exhaust their group shares. However, the amount of the delay cannot be defined in the input model. Rather, it has to be determined based on the system performance. This section describes how the utilization and waiting time results are used to find the demand for a delay used to throttle a group. The section that follows describes how these calculations are integrated into the analytic solver LQNS.

4.1 Mappings to Dynamic Parameters

One necessary step for an analytic solution for a FS scheduler is to find suitable expressions for mapping output parameters such as throughputs and utilizations to input parameters. The DPS functions were developed based on the ideas of the CFS scheduler used in [14], Sect. 3.2. The time that a group spends waiting is used to handle the case that the group gains the FS processor after it exhausts its share. Inside the LQNS solver, the extra delays of groups at a FS processor are modelled by surrogate delays [12].

The extra delay of each group becomes a dynamic parameter, D_i which is used as the service time of a new *delay task* in the model. When the utilization of a group i at the FS processor U_i is greater than its group share GS_i, the usage of this group is reduced by increasing the time the group spends at its delay task. The substitution function for the dynamic parameter D_i then becomes (2) when U_i is greater than GS_i, and (3) when U_i is smaller than GS_i.

$$
\begin{aligned}
\Delta W_i &= \frac{(U_i - GS_i)}{U_i} \times W_i \\
\$D_i^{(n+1)} &= \$D_i^{(n)} + \Delta W_i
\end{aligned} \quad (U_i > GS_i) \tag{2}
$$

$$
\begin{aligned}
\Delta W_i &= \min(\frac{GS_i - U_i}{U_i} \times W_i, \$D_i^{(n)}) \\
\$D_i^{(n+1)} &= \$D_i^{(n)} - \frac{\Delta W_i}{2};
\end{aligned} \quad (U_i < GS_i) \tag{3}
$$

where: W_i and U_i are the mean residence time and processor utilization of a group i at the FS processor.

4.2   Redistributing Surplus Shares

If the FS processor has some spare capacity, a group with a guarantee share may
have a chance to consume some of the extra CPU time. The number of extra
shares a group can get is proportional to its original share using (4).

$$
\begin{aligned}
GS^- &= \sum_i^{i \in \mathcal{P}} (GS_i - U_i) \\
GS^+ &= \sum_i^{i \in \mathcal{Q}} GS_{(g,i)} \\
GS'_{(g,i)} &= GS_{(g,i)} \times \left(1 + \frac{GS^-}{GS^+}\right)
\end{aligned}
\tag{4}
$$

where: GS^- is the total surplus shares that can be redistributed, GS^+ is the
total guarantee shares that can be increased, $\mathcal{P}$ is a set of groups that are able to
contribute their spare shares, and $\mathcal{Q}$ is a set of groups that may receive additional
shares.

For any groups in the set $\mathcal{P}$, their group shares can be either guarantees or
caps while all groups in the set $\mathcal{Q}$ have guarantee shares only. Also, all the groups
in set $\mathcal{P}$ have some spare shares and must not have any delay at the delay server
($\$D_{i \in \mathcal{P}} = 0$). Conversely, all the groups in set $\mathcal{Q}$ must have some delays at the
delay server, which indicates these groups definitely need more resources.

Listing 1.1. LQX function for surplus share redistribution

```
1   function share_redistribution( ngroups, groups[], group_delay[])
2   {
3       new_share[ngroups];
4       share_can_increase = 0.0;
5       surplus_share = 0.0 ;
6       can_increase[ngroups];        /* Boolean */
7       for(int g=0; g<ngroups; g++){
8           can_increase[g]=False;
9           new_share[g]=groups[g].share;
10          if ((groups[g].utilization <groups[g].share)
11                          && (group_delay[g]<=0)){
12              surplus_share+=groups[g].share-groups[g].utilization;
13          }else if ((groups[g].utilization>groups[g].share)
14                          && !(groups[g].isCap) ){
15              share_can_increase +=  groups.share;
16              can_increase[g] = True;
17          }
18      }
19      if (share_can_increase > 0.05 ){
20          unit_share = surplus_share /share_can_increase;
21          for(int g=0; g<ngroups; g++){
22              if( can_increase[g] ){
23                  new_share[g] = groups[g].share *(1+ unit_share);
24                  new_share = (new_share >1)? 1.0 :  new_share;
25              }
26          }
27      }
28      return (new_share);
29  }
```

Equation (4) gives the total surplus shares that can be redistributed, which is the total of the differences between group utilization and group shares of all the groups in the set $\mathcal{P}$, while (4) gives the total guarantee shares of all the groups in set $\mathcal{Q}$. Therefore, the fraction of $\frac{GS^-}{GS^+}$ in (4) defines how many extra shares a group may receive.

The surplus share is redistributed by using Listing 1.1. The condition at line 19 in Listing 1.1 is intended to reduce the total number of iterations.

5 Integrating DPS into LQNS

Layered queueing networks are solved analytically by LQNS grouping tasks and processors into submodels which are each solved using conventional MVA. The results from one submodel are used to set parameters in the other submodels. The solution iterates among the submodels until the change in the parameters becomes sufficiently small.

Dynamic Parameter Substitution takes this approach one step further by applying a function to the output of one submodel then using its results as input into a different submodel. For FS scheduling, two extensions are made to the LQN analytic solution:

1. A new type of LQN submodel, *FairShareSubmodel* was created as a subclass of MVASubmodel and used to handle the DPS iterations.
2. A new type of queueing station, *FairShareServer* was created as a subclass of FCFS queue, and used to handle the priority issue of groups.

5.1 FairShareSubmodel

A *FairShareSubmodel* contains at least one FS processor. The solution of *FairShareSubmodel*, shown in Listing 1.2, consists two levels of iterations: DPS iterations and MVA iterations. The MVA iterations are used to solve the submodel and are invoked during every DPS iteration.

In each DPS iteration, *FairShareSubmodel* performs the substitutions of dynamic parameters using the substitution functions (2) and (3). The new values are used in the following MVA iterations. The results of an MVA solution, including the utilization of all the groups, are saved and are fed into the substitution functions to find out the extra delays of the groups in the next iteration. The DPS iterations stop when the DPS convergence test is satisfied.

If the values of the dynamic parameters are varied in every DPS iteration, the LQNS-DPS approach may lead to convergence problems. To avoid this issue, the DPS iterations are managed using the algorithm shown in Listing 1.3. This algorithm is grouped into five steps:

Listing 1.2. FairShareSubmodel Solution Algorithm.

```
1   begin
2       reset groups;
3       do:
4           LQNS-DPS();
5           Solve FairShareSubmodel s using MVA;
6           Save group utilization;
7       until{ DPS convergence}
8       Save waiting times for FairShareSubmodel s;
9       Set think times for submodel s+1;
10  end
```

1. Step 1 is the first DPS iteration. In this step, the submodel is solved like a regular MVASubmodel without any added-on extra delays. The results of this step, determine which groups need to be adjusted because they used more resource than they are entitled to.
2. Step 2 is the substitution step. The DPS substitution functions are invoked prior to the MVA iterations. This step stops when the utilizations of all the groups are not greater than their group share (regardless of guarantees or caps). The results of this step give the group utilizations when all groups are capped. On the other hand, the results also show whether surplus shares exist. If no surplus share exists, this is a fully utilized resource, then the DPS iterations enter Step 5. Otherwise, DPS iterations enter Step 3.
3. Step 3 is for redistributing surplus shares. The redistribution algorithm for surplus shares is in Listing 1.1. The results of step 3 determine how much extra resource a group with a guarantee share may obtain. Then DPS iterations enter Step 4.
4. Step 4 is also a substitution step, which is almost same as Step 2, except the enlarged group shares are used in the substitution functions. The results of Step 4 provide the group utilizations which satisfy the increased group shares. At the end of Step 4, if there still are some surplus shares remaining, the DPS iterations go back Step 3 again. Otherwise, the DPS iterations enter Step 5.
5. Step 5 is the final step. In this step, all the DPS operations are disabled, and all the dynamic parameters are not changed. *FairShareSubmodel* is again solved as a regular MVASubmodel. The purpose of this step is to assure the convergence of the entire model.

5.2   FairShareServer

FairShareServer is a new type of queue and used to handle the priority issue of 'contributing' groups at a FS processor. It is a subclass of *FCFSServer* (a FCFS queue) in the LQNS solver. The characteristics of a *FairShareServer* are:

1. *FairShareServer* is a FCFS queue when *NO* group is contributing its extra share.

Listing 1.3. LQNS-DPS algorithm

```
1   begin
2      if (step != 5):
3         reset_groups();
4      long dps_iter = 0;   /* initialize */
5      double dps_delta = 0.;
6      bool _dps_converage = false;
7      do:
8         step = check_dps_step();
9         dps_iter += 1;
10        if (step == 2 or step == 4):
11           dps_delta += group_adjust();
12        else if (step == 3):
13           redistribute_share();
14           group_reset();
15           dps_delta += group_adjust();
16        endif
17        MVA_solve(); /* solve submodel using MVA iterations */
18        if (step == 1):
19           save first_step_util;
20        else if (step == 5):
21           _dps_converage = true; /* stops the DPS iteration */
22        else:
23           _dps_converage = dps_converge_test( dps_delta );
24        endif
25     until{ _dps_converge}
26     save MVA results;
27     save group utilization;
28  end
```

2. *FairShareServer* is a priority queue with preemption when one or more groups are contributing their unused shares, *AND* some groups are consuming these extra shares.

In the first case, all groups have the same priority when they queue at the resource, therefore, tasks from any groups are served as a FCFS order. In the second case, some surplus shares exist. The groups that are consuming the extra share must have extra arrivals at the FS processor to use the remaining available resource. The extra arrivals may increase the queueing delays of the requests from the contributing groups. Therefore, to eliminate extra queueing delays, the requests from the groups contributing the extra share must have a higher priority than the requests from the groups receiving the extra share. If a request (from the contributing groups) arrives at the resource after these extra arrivals (from the consuming groups), the request will preempt these customers that are in front of it.

In the LQN analytic solution, handling group priority is realized by applying the 'maximum queueing delays' to the contributing groups. In fact, the queueing delays occurred when all groups are capped are the maximum queueing delays of all the contributing groups. These maximum queueing delays can be found at the end of Step 2 of the DPS iterations. These maximum queueing delays are passed into MVA iterations in Step 4 of the DPS iterations as the bounds of the queueing delays to limit the increment of the residence time of the requests from the contribution groups.

5.3   Convergence Control

LQNS-DPS has two levels of convergence tests. First, the convergence criteria for both Step 2 and Step 4 is that the sum of the squares of the difference of utilizations $\Delta\$U_i$ for all groups between two successive **DPS iterations** is less than 5×10^{-5}. Second, adding Step 5 ensures the convergence of entire model. All the dynamic parameters are using finalized values. This step can be reached from Step 2 and Step 4 after they converge. The condition for entering Step 5 is that the sum of the squares of $\Delta\$U_i$ for all groups between two successive **submodel iterations** is less than 5×10^{-5}.

6   Verification of the DPS Approach

In order to verify the LQNS-DPS approach, two sets of models are used in the section. First, a simplified model with a FS processor is used to verify that the LQNS-DPS approach can handle all the cases of fair share scheduling. The second set models are randomly generated models, with a larger size and higher complexity. The results found by DPS are compared with the results of simulations, which run in the confidence interval of 95 %, ±2.0 %.

6.1   A Simplified Model

This model, shown in Fig. 2, contains three classes of clients, $c1$, $c2$ and $c3$, which make requests to one of server tasks $s1$, $s2$ and $s3$. These server tasks are running at a FS processor *ApplicCPU*. Task $s1$ belongs to *Group1* while tasks $s2$ and $s3$ belong to the other group, *Group2*.

In the experiments, the group share of Group1 is varied from 0.1 to 0.9 and Group2 is from 0.9 to 0.1. The model was solved multiple times under the following four configurations: (a) both groups have cap shares; (b) both groups have guarantee shares; (c) Group 1 has a cap share and Group 2 has a guarantee share; (d) opposite to (c) only Group 2 is capped.

The results for these four configurations by solving from the LQNS-DPS approach and simulation are shown in Fig. 3. The chosen workload of Group 1 can only consume a maximum of the 60 % of the processor resource, while Group 2 always consumes all the available resource. In configurations (a) and (d), Group 2 is prevented from consuming the spare resource since it is capped. Therefore, a larger group share for Group 1 starves the FS processor. On the other hand, in configurations (b) and (c), the FS processor has surplus shares when Group 1 has a share more than 0.5. These surplus shares are consumed by

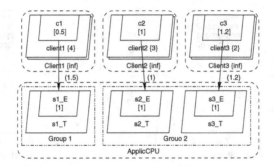

Fig. 2. A test model with 3 tasks in two groups.

Group 2, then the FS processor becomes fully utilized again. The experiments for this model are able to cover all the situations of the competitions between these two groups at the FS processor. Without share redistribution (a and d), the mean error rates of LQNS-DPS are less than 1 %, while in the cases with share redistribution (b and c), the mean error rates are around 5 %.

The execution times of the LQNS-DPS and simulation are shown in Table 1 and clearly show that the analytic method, LQNS-DPS, is significantly faster than simulation.

Table 1. The execution time of the LQNS-DPS and simulations.

Type of group shares	LQNS-DPS execution (s)	LQSIM execution (s)
Both cap shares	0.012	4.236
Both guarantee	0.010	3.667
Group 1 capped	0.011	3.346
Group 2 capped	0.009	3.787

6.2   Randomized Model

This random model was generated by lqngen [9], which is one of LQN tools. Using lqngen, random models can be generated by specifying the number of layers, the number of reference tasks, the total number of tasks or processors, as well as the shape of system models, such as pyramid (more servers at lower layers), funnel etc.

The random model, which is shown in Fig. 1, has three reference tasks ($c0$, $c1$ and $c2$) and eight server tasks ($t0\sim t7$), which are located in three layers. These server tasks are hosted by two processors, $p0$ and $p1$. The processor $p1$ is a FS processor and hosts six server tasks, $t0$, $t3$, $t5$, $t7$, and $t1$, $t2$, which are in the two groups, $Group1$ and $Group2$. Two sets of experiments were performed and all the input parameters are also randomized in the input file of this random model through LQX [17].

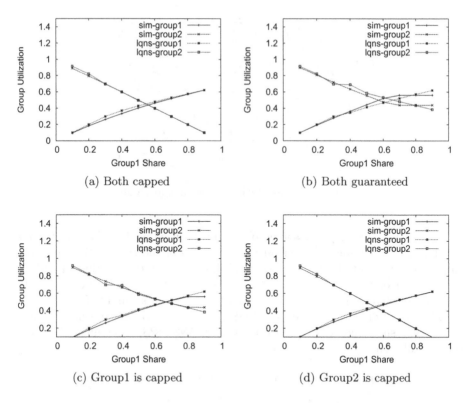

Fig. 3. Results from LQNS-DPS and LQSIM.

In the first set of experiments, both the two groups are capped using two pairs of group shares (0.6, 0.4) and (0.2, 0.8), and 50 sets of random parameters for each pair of group share were performed. Table 2 show the difference of the

Table 2. Results for the randomized model.

%Δ	Group1					Group2			Processor	Throughput		
	U_{t0}	U_{t3}	U_{t5}	U_{t7}	U_{Total}	U_{t1}	U_{t2}	U_{Total}	U_{proc}	λ_{C0}	λ_{C1}	λ_{C2}
$CS = 0.6, CS = 0.4$												
mean	−0.20	−2.16	−2.15	−2.16	−1.61	3.41	3.49	4.91	0.84	−0.15	3.29	3.53
sdev	3.55	3.18	3.15	3.19	3.07	13.20	6.39	7.41	3.52	3.57	13.21	6.33
$CS = 0.2, CS = 0.8$												
mean	3.92	2.11	2.24	2.16	2.65	3.23	2.53	2.80	2.72	3.98	3.22	2.55
sdev	5.13	5.22	5.19	5.25	5.10	1.41	0.87	0.59	1.10	5.11	1.35	0.91
$GS = 0.2, CS = 0.8$												
mean	−0.47	−2.49	−2.38	−2.45	−1.93	5.87	1.31	2.25	1.17	−0.56	5.86	1.31
sdev	7.07	6.85	6.85	6.84	6.85	3.09	1.11	0.95	1.68	7.10	3.05	1.12

Table 3. The execution time of the LQNS-DPS and simulations.

Group shares	LQNS-DPS (Sec.)	LQSIM (Sec.)
$(CS_{g1} = 0.6, CS_{g2} = 0.4)$	0.1061	33.827
$(CS_{g1} = 0.2, CS_{g2} = 0.8)$	0.1178	47.969
$(CS_{g1} = 0.2, GS_{g2} = 0.8)$	0.1173	43.689

results of LQNS-DPS compared to the simulation results. All the simulation results found by LQSIM, are with a confidence interval of $95\% \pm 2.0\%$. The values in the row labelled "mean" are the mean values of the difference between LQNS-DPS and LQSIM; while the values in the row of "sdev" are the standard deviations of these differences.

The second set experiments, only *Group2* is capped at 0.8, so *Group1* has a guarantee share of 0.2; the mean number of customers in *Group2* is reduced. Therefore, *Group2* may not have enough workload to consume its large share of 0.8, so a spare share appears and can be consumed by *Group1*.

Table 2 also shows the mean error and standard deviation for comparing LQNS-DPS and simulation over 50 different sets of parameters. The mean error rates of the group utilizations and the processor utilization are around 2%. The maximum mean error rate of task utilizations is 5.79 only at task $t1$, and for all the other tasks, the mean error rates are around $2\sim3\%$.

The computational efforts for LQNS-DPS and simulation (the average execution times of these experiments on randomized models) are shown in Table 3. Again, the analytic method, LQNS-DPS, is 400 times faster than simulations.

7   Conclusions

Previously, LQN models with processors with FS scheduling were only handled using the LQN simulator. This paper revisits these LQN models and solves them analytically by integrating the method of dynamic parameter substitution into the LQN analytic solution directly. Dynamic Parameter Substitution extends the solution of LQN models through decomposition and iteration by incorporating more advanced transformations of the results of one submodel into the parameters of another. This solution produced good accuracy compared to simulation with both cap and guarantee shares, and in the presence of surplus share for models with guarantee shares. The mean error in utilization of any task or group was never greater than 6% when compared to simulation.

The implementation was realized using a new *FairShareSubmodel* and *FairShareServer* queue within the analytic LQN Solver. Using a FairShareServer within a FairShareSubmodel limits the iterations needed to handle FS scheduling to a single submodel, thus reducing overhead and speeding convergence. Furthermore, this approach allows the priorities of the groups sharing the processor to vary so that models with surplus capacity and guarantee shares can be solved accurately. Finally, the approach is scalable and was found to be 400 times faster than simulations producing confidence intervals of $\pm 2\%, 95\%$ of the time.

References

1. Bolker, E., Ding, Y.: On the performance impact of fair share scheduling. In: CMG Conference, Orlando, FL, USA, 10–15 December 2000, vol. 1, pp. 71–82. Computer Measurement Group (2000)
2. Bolker, E., Ding, Y., Rikun, A.: Fair share modeling for large systems: aggregation, hierarchical decomposition and randomization. In: CMG Conference, Anaheim, CA, USA, vol. 2, pp. 807–818. Computer Measurement Group (2001)
3. Chandra, A., Adler, M., Goyal, P., Shenoy, P.: Surplus fair scheduling: a proportional-share CPU scheduling algorithm for symmetric multiprocessors. In: Proceedings of the 4th Symposium on Operating System Design and Implementation (OSDI 2000), San Diego, California, 23–25 October 2000, pp. 45–58 (2000)
4. Cherkasova, L., Gupta, D., Vahdat, A.: Comparison of the three CPU schedulers in Xen. SIGMETRICS Perform. Eval. Rev. **35**(2), 42–51 (2007)
5. Demers, A., Keshav, S., Shenker, S.: Analysis and simulation of a fair queueing algorithm. In: Symposium Proceedings on Communications Architectures & Protocols, SIGCOMM 1989, pp. 1–12. ACM, New York (1989)
6. Duda, K.J., Cheriton, D.R.: Borrowed-virtual-time (BVT) scheduling: supporting latency-sensitive threads in a general-purpose scheduler. SIGOPS Oper. Syst. Rev. **33**(5), 261–276 (1999)
7. Figueiredo, R.J., Dinda, P., Fortes, J., et al.: A case for grid computing on virtual machines. In: Proceedings of the 23rd International Conference on Distributed Computing Systems, pp. 550–559, May 2003
8. Franks, G., Al-Omari, T., Woodside, M., Das, O., Derisavi, S.: Enhanced modeling and solution of layered queueing networks. IEEE Trans. Softw. Eng. **35**(2), 148–161 (2009)
9. Franks, G., Maly, P., Woodside, M., Petriu, D.C., Hubbard, A.: Layered Queueing Network Solver and Simulator User Manual. Real-time and Distributed Systems Lab, Carleton University, Ottawa (2014)
10. Ghodsi, A., Sekar, V., Zaharia, M., Stoica, I.: Multi-resource fair queueing for packet processing. ACM SIGCOMM Comput. Commun. Rev. **42**(4), 1–12 (2012)
11. Henry, G.J.: The UNIX system: the fair share scheduler. AT&T Bell Lab. Tech. J. **63**(8), 1845–1857 (1984)
12. Jacobson, P.A., Lazowska, E.D.: Analyzing queueing networks with simultaneous resource possession. Commun. ACM **25**(2), 142–151 (1982)
13. Kay, J., Lauder, P.: A fair share scheduler. Commun. ACM **31**(1), 44–55 (1988)
14. Li, L., Franks, G.: Performance modeling of systems using fair share scheduling with layered queueing networks. In: Proceedings of the 17th IEEE/ACM International Symposium on Modeling, Analysis and Simulation of Computer and Telecommunications Systems (MASCOTS 2009), London, 21–23 September 2009, pp. 1–10 (2009)
15. Li, L., Franks, G.: Hybrid performance modeling using layered queueing networks. In: CMG-2012, Las Vegas, Nevada, USA, 3–7 December 2012, vol. 1, pp. 85–93. Computer Measurement Group, Curran Associates Inc. (2012)
16. Li, L., Franks, G.: Improved interlock correction when solving layered queueing networks using decomposition. In: 2015 IEEE 28th Canadian Conference on Electrical and Computer Engineering (CCECE), pp. 541–546, May 2015

17. Mroz, M., Franks, G.: A performance experiment system supporting fast mapping of system issues. In: Proceedings of the 4th International ICST Conference on Performance Evaluation Methodologies and Tools (VALUETOOLS 2009), Pisa, Italy, 20–22 October 2009, pp. 246–255 (2009)

18. Nagle, J.: On packet switches with infinite storage. IEEE Trans. Commun. **35**(4), 435–438 (1987)

19. Neilson, J.E.: PARASOL Users Manual. School of Computer Science, Carleton University, Ottawa, Canada, (version 3.1, edn.)

20. Pabla, C.S.: Completely fair scheduler. Linux J. **2009**(184), 4 (2009)

21. Ramanathan, P., Agrawal, P.: Adapting packet fair queueing algorithms to wireless networks. In: Proceedings of the 4th Annual ACM/IEEE International Conference on Mobile Computing and Networking, MobiCom 1998, pp. 1–9. ACM, New York (1998)

22. Stiliadis, D., Varma, A.: Efficient fair queueing algorithms for packet-switched networks. IEEE/ACM Trans. Netw. **6**(2), 175–185 (1998)

23. Waldspurger, C.A., Weihl, W.E.: Lottery scheduling: flexible proportional-share resource management. In: Proceedings of the 1st USENIX Conference on Operating Systems Design and Implementation (OSDI 1994), 14–17 November 1994, p. 1. USENIX Association, Berkeley (1994)

24. Waldspurger, C.A., Weihl, W.E.: Stride scheduling: deterministic proportional-share resource management. Technical report TM-528, Massachusetts Institute of Technology, Cambridge, MA, USA, June 1995

25. Weng, C., Wang, Z., Li, M., Lu, X.: The hybrid scheduling framework for virtual machine systems. In: Proceedings of the 2009 ACM SIGPLAN/SIGOPS International Conference on Virtual Execution Environments, VEE 2009, pp. 111–120. ACM, New York (2009)

26. Wood, T., Shenoy, P.J., Venkataramani, A., Yousif, M.S.: Black-box and gray-box strategies for virtual machine migration. In: NSDI, vol. 7, p. 17 (2007)

27. Woodside, C.M.: Controllability of computer performance tradeoffs obtained using controlled-share queue schedulers. IEEE Trans. Softw. Eng. **12**(10), 1041–1048 (1986)

Concentrated Matrix Exponential Distributions

Illés Horváth[1]([✉]), Orsolya Sáfár[2], Miklós Telek[3], and Bence Zámbó[4]

[1] MTA-BME Information Systems Research Group, Budapest, Hungary
[2] Department of Analysis, Budapest University of Technology and Economics,
Budapest, Hungary
pollux@math.bme.hu
[3] Department of Networked Systems and Services, Budapest University
of Technology and Economics, Budapest, Hungary
[4] Institute of Mathematics, Budapest University of Technology and Economics,
Budapest, Hungary

Abstract. We revisit earlier attempts for finding matrix exponential (ME) distributions of a given order with low coefficient of variation (cv). While there is a long standing conjecture that for the first non-trivial order, which is order 3, the cv cannot be less than 0.200902 but the proof of this conjecture is still missing.

In previous literature ME distributions with low cv are obtained from special subclasses of ME distributions (for odd and even orders), which are conjectured to contain the ME distribution with minimal cv. The numerical search for the extreme distribution in the special ME subclasses is easier for odd orders and previously computed for orders up to 15. The numerical treatment of the special subclass of the even orders is much harder and extreme distribution had been found only for order 4.

In this work, we further extend the numerical optimization for subclasses of odd orders (up to order 47), and also for subclasses of even order (up to order 14). We also determine the parameters of the extreme distributions, and compare the properties of the optimal ME distributions for odd and even order.

Finally, based on the numerical results we draw conclusions on both, the behavior of the odd and the even orders.

Keywords: Matrix exponential distributions · Minimal coefficient of variation

1 Introduction

The class of matrix exponential (ME) distributions, along with the subclass of phase type (PH) distributions are widely used for a number of reasons. The class PH is widespread in Markovian modeling: it can be efficiently used for approximating non-Markovian stochastic models (where random durations are non-exponential) by Markovian models, while maintaining a simple analytic formula [2].

The class ME exhibits an even wider range of behavior, allowing for even more efficient approximations; the cost is that a simple stochastic interpretation with

© Springer International Publishing AG 2016
D. Fiems et al. (Eds.): EPEW 2016, LNCS 9951, pp. 18–31, 2016.
DOI: 10.1007/978-3-319-46433-6_2

background Markov chain is no longer available. Still, simple analytic formulas are retained, so it is still useful for efficient calculations.

The class PH is known to approximate deterministic delay poorly. That is why it is an important application of either class to approximate the deterministic distribution; in other words, we are looking for PH and ME distributions which are highly concentrated. A usual measure of concentration is the coefficient of variation (cv). For the class PH, it is known that the minimal cv depends only on the order n, it is $1/n$ and is obtained by the Erlang distribution [4].

However, for the class ME, only partial results are available. It has been known that ME(2) $\equiv$ PH(2) [12], that is the class of order 2 ME distributions is identical with the class of order 2 PH distributions.

For higher order ME, numerical investigations indicate that the minimal cv can be significantly lower compared to the class PH of the same order [5,10]. No analytic results are available for the minimal value of cv in the class ME(n) for any $n \geq 3$. [5] lists some conjectures for the minimal value of cv; later, [10] numerically optimizes cv for a convenient subclass of ME(n) for odd values of n up to $n \leq 15$ and even values of n up to $n \leq 4$.

In the present paper, we extend the previously used approaches to larger values of n, for both odd and even orders and based on the numerical results we draw conclusions on the trends of their behavior with increasing orders.

The rest of the paper is organized as follows. In Sect. 2, we give the necessary mathematical background and notations are introduced. In Sect. 3, we introduce special subsets of ME(n) distributions for both odd and even values of n, detail the techniques and calculations necessary for finding distribution with minimal cv in the respective subsets, and present the results of the numerical optimization (some of which are put in the Appendix for better readability).

2   Preliminaries

Definition 1. *Let X be a non-negative continuous random variable with cumulative distribution function (cdf)*

$$F_X(t) = \Pr(X < t) = 1 - \alpha e^{\mathbf{A}t}\mathbf{1}, \quad t \geq 0 \tag{1}$$

where α is a row vector of length n, $\mathbf{A}$ is a matrix of size $n \times n$ and $\mathbf{1}$ is a column vector of ones of size n. Then X is matrix exponentially distributed with representation $(\alpha, \mathbf{A})$, or shortly X is ME($\alpha, \mathbf{A}$) distributed, where α is referred to as the initial vector and n as the order.

The probability density function (pdf) of X is then

$$f_X(t) = -\alpha \mathbf{A} e^{\mathbf{A}t}\mathbf{1}, \quad t \geq 0. \tag{2}$$

We note that, as the above terminology suggests, α and $\mathbf{A}$ is not unique to F_X; the same function F_X may have several different α, A pairs referred to as representations. Further more, not even the order n of the representation is unique. There are $ME(\alpha_1, \mathbf{A}_1)$ of order m and $ME(\alpha_2, \mathbf{A}_2)$ of order n such that $m \neq n$ and $ME(\alpha_1, \mathbf{A}_1)$ and $ME(\alpha_2, \mathbf{A}_2)$ have the same distribution function, see e.g. [3].

Definition 2. *The class ME(n) contains matrix exponential distributions which have a representation of order at most n.*

Definition 3. *If X is $ME(\alpha, \mathbf{A})$-distributed, and α and $\mathbf{A}$ satisfies the following assumptions:*

- $\alpha_i \geq 0$,
- $A_{i,j} \geq 0$ for $i \neq j$, $A_{j,j} \leq 0$,
- $A\mathbf{1} \leq 0$

then we say X is phase type (PH) distributed, or shortly $PH(\alpha, \mathbf{A})$ distributed. A representation $(\alpha, \mathbf{A})$ satisfying the above properties is called Markovian.

Definition 4. *The class PH(n) contains matrix exponential distributions which have a Markovian representation of order at most n.*

Based on the Jordan decomposition of $\mathbf{A}$ in (2) the probability density function (pdf) of an ME distribution has the following general form:

$$f(t) = \sum_{i=1}^{k} \sum_{j=0}^{N_i - 1} c_{i,j} t^j e^{\lambda_i t}, \tag{3}$$

where $\lambda_1, \ldots, \lambda_k$ are eigenvalues of $\mathbf{A}$, and λ_i has multiplicity N_i. Some of the eigenvalues may be complex (in which case they come in complex conjugate pairs). Some eigenvalues of $\mathbf{A}$ may not be present in f; the eigenvalues which are present in the pdf are referred to as *contributing* eigenvalues. All contributing eigenvalues must have negative real parts, and among the contributing eigenvalues there is always a dominant real eigenvalue with maximal real part [6,8]). λ_1 denotes this dominant eigenvalue. In the class ME, there may be complex eigenvalues with real part equal to λ_1; in the class PH, this is not possible. Further differences between the classes ME and PH can be found in, e.g., [6,7,9].

To keep the subsequent discussion simple, we sometimes calculate with *unnormalized* pdfs, that is, $\int_0^\infty f(t)\mathrm{d}t = 1$ is not required, only

$$0 < \int_0^\infty f(t)\mathrm{d}t = m_0 < \infty.$$

Of course, this means that $\frac{f(t)}{m_0}$ is the proper normalized pdf corresponding to X. Then the moments can be calculated from an unnormalized f as

$$\mathbf{E}(X^n) = \frac{m_n}{m_0},$$

where

$$m_n = \int_0^\infty t^n f(t)\mathrm{d}t.$$

Definition 5. *The coefficient of variation (cv) of $X \sim ME(\alpha, \mathbf{A})$ is*

$$cv(X) = \frac{\mathbf{E}(X^2) - (\mathbf{E}(X))^2}{(\mathbf{E}(X))^2} = \frac{m_2 m_0}{m_1^2} - 1.$$

The notation cv(f) will be used as well.

The coefficient of variation is a widely used measure of probability concentration of positive random variables. It is invariant to scaling of the variable (e.g., multiplying with a positive number, or changing the unit the random variable is expressed in); that is, $ME(\alpha, \mathbf{A})$ and $ME(\alpha, c\mathbf{A})$ (or, correspondingly, $f(t)$ and $f(t/c)/c$) have the same cv. This property also allows us to conveniently scale the considered distributions; for example, the dominant eigenvalue may be assumed to be -1.

For the class PH(n), the minimal cv is known.

Theorem 1. *[4] For $X \in PH(n)$, $cv(X) \geq \frac{1}{n}$, and the minimum is obtained for the Erlang distribution with parameters (n, λ) where $\lambda > 0$ is arbitrary.*

We note that in accordance with our previous remark on scaling, λ (the dominant eigenvalue) does not affect the minimal cv.

However, an analytical result similar to Theorem 1 for $\mathrm{argmin}\{cv(f) : f \in ME(n)\}$ is available only for $n \leq 2$. $ME(1) = PH(1)$ is just the family of exponential distributions with cv $= 1$, while $ME(2) = PH(2)$ with cv $\geq 1/2$. For $n \geq 3$, [10] numerically optimizes cv for a convenient subclass of $ME(n)$ for odd values of the order n up to $n = 15$, and conjectures that the minimal value of cv is indeed obtained within the given subclass.

3   Numerical Optimization of cv for ME(n)

3.1   Optimization for Odd n

Following [5,10], for odd n, we look to minimize coefficient of variation in the subclass containing ME probability distribution functions of the following form (unnormalized pdf):

$$f(t) = e^{-t} \prod_{i=0}^{(n-1)/2} \cos^2(\omega t - \phi_i). \tag{4}$$

This is the same subclass presented in [10], where the technology for computing the optimal parameters of (4) is the following: the moments and cv can be calculated analytically by Laplace-transform with Mathematica, then a numerical optimization is carried out for the variables $\omega, \phi_1, \ldots, \phi_{(n-1)/2}$. For more details, see [10]. The main difficulty for this approach is that as n increases, the analytical expression for cv gets more difficult to compute.

In the present paper, we use a different approach with only numerical steps. The moments are calculated by numerical integration for specific values of

$\omega, \phi_1, \ldots, \phi_{(n-1)/2}$. We found the minimum implementing an evolution strategy. Starting with one feasible solution, the parameter values were changed by adding a normally distributed random value for each parameter (mutating the solution). The deviation was changed according to Rechenberg's 1/5 rule; if the success ratio (that is, the cv has decreased for the new parameter values) after 20 steps is over 1/5, then the deviation is sligthly increased, in order to ensure the exploration of the search space. If the ratio is below 1/5 then the deviation is decreased in order to get close to the local optimum. Due to the numerous local optima and the high number of parameter values to be optimised, a population size of one was used, exploring the search space by the self-adaptation of the deviation. We changed to the new modified parameter values only if the cv decreased, otherwise we tried a new mutation. The best result from several runs was selected for each n. See [11] for more on the theory and [1] for the code in Matlab.

This approach is feasible for significantly higher values of n. We calculate the optimum for up to $n \leq 47$ and we also list the argmin parameters.

Non-negativity of $f(t)$ for $t \geq 0$ in (4) is guaranteed from its form in (4). The structure of the pdf is the following: the part $\prod_{i=0}^{(n-1)/2} \cos^2(\omega t - \phi_i)$ is periodic

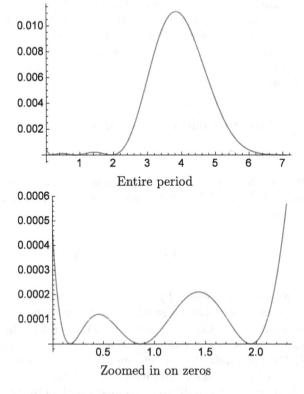

Fig. 1. Optimal f for $n = 7$

with period ω/π, and has $(n-1)/2$ zeros within a period. The exponential term in (4) introduces a decay that renders the part of the pdf from the second period on negligible. Within the first period, the optimal arrangement of zeros is best seen the following: the zeros are roughly (but not exactly) equidistant, starting from slightly above 0, but leaving a large gap between the last zero and the end of the period; see Fig. 1. The closely situated zeros bring the pdf close to 0, while the large gap results in a relatively large, concentrated bump.

The form of (4) is yet again slightly different from the form given in (2) or (3); in Appendix A it is shown how (4) can be converted back to (3).

Table 2 contains the parameters which provide the minimal cv in $ME(n)$. Due to readability, only values for $n \leq 17$ are displayed here; the rest of the list of optimal parameter values for n up to 47 can be found in Table 4 in Appendix B. Table 1 contains the minimal values of cv for various values of n; even values of n are also included in the table for easier comparison. Note that for odd n

Table 1. Minimal values of cv for various values of n

n	cv	1/cv	n	cv	1/cv
3	0.20090	4.9776	4	0.14981	6.6752
5	0.081264	12.306	6	0.075532	13.2394
7	0.042880	23.321	8	0.041349	24.1845
9	0.026157	38.231	10	0.025589	39.079
11	0.017494	57.163	12	0.017237	58.015
13	0.012470	80.195	14	0.012337	81.060
15	0.0093128	107.38			
17	0.0072074	138.75			
19	0.0057368	174.31			
21	0.0046708	214.10			
23	0.0038745	258.10			
25	0.0032646	306.31			
27	0.0027874	358.75			
29	0.0024053	415.76			
31	0.0020760	481.70			
33	0.0018094	552.66			
35	0.0015907	628.64			
37	0.0014092	709.64			
39	0.0012568	795.66			
41	0.0011278	886.71			
43	0.0010051	994.94			
45	0.00088322	1132.2			
47	0.00078490	1274.0			

Table 2. Optimal parameter values for odd $n \leq 17$

Order	ω	$\phi_1, \phi_2, \phi_3, \ldots$
3	1.03593	0.337037
5	0.474055	1.67698; 2.10333
7	0.442459	1.64632; 1.95221; 2.4311
9	0.418775	1.62842; 1.86379; 2.23549; 2.69568
11	0.400272	1.61684; 1.80633; 2.10839; 2.48269; 2.90758
13	0.385334	1.60882; 1.7663; 2.01958; 2.33446; 2.69071; 3.0794
15	0.372959	0.07936; 1.60297; 1.73697; 1.95429; 2.22554; 2.53226; 2.86521
17	0.362491	0.19780; 1.59854; 1.90442; 1.71468; 2.14228; 2.41145; 2.70297 3.012594372647776

(even $n + 1$), the difference between the optimal cv for n and $n + 1$ is rather small, especially for larger values of n. The optimum in ME(n) and ME($n + 1$) is further compared in Sect. 3.2 in more detail, and a detailed analysis of even values of n follows in Subsect. 3.3.

3.2   ME(n) for Even n

For even values of n, we look to minimize coefficient of variation in the subclass containing ME probability distribution functions of the following form:

$$f(t) = a_3 e^{t\lambda_2} + e^{-t}\left(a_2 + \prod_{i=0}^{(n-2)/2} \cos^2(\omega t - \phi_i)\right) \tag{5}$$

Again, this subclass was already present in [5] but no optimization was carried out except for $n = 4$.

We present an intuitive argument that explains the subclass (5) and sheds some light on the difference between the optimal function in ME(n) and ME($n+1$) for odd n.

Compared to (4), the novelty in (5) is the term $(a_3 e^{t\lambda_2} + a_2 e^{-t})$. To understand the formulas better, we note that in (5),

$$\prod_{i=0}^{(n-1)/2} \cos^2(\omega t - \phi_i) \tag{6}$$

is periodic with period π/ω, and it has $(n - 1)/2$ zeros for each period. Each zero effectively serves to make the function (6) more concentrated by bringing it close to 0 on a large portion of the interval $[0, \pi/\omega]$ (see Fig. 1), while the exponential term e^{-t} in (4) brings a decay that renders the concentrated parts from the second interval on irrelevant.

Now for $n + 1$, a new zero within the period is not possible, as an additional $\cos^2$ term would correspond to $n + 2$ in the order. So instead we can use it to change some other part of the function instead. To have the most effect on the value of cv, we use this additional term $(a_3 e^{t\lambda_2} + a_2 e^{-t})$ to decrease the value of the function around 0 to as small as possible. This leads naturally to the assumption $f(0) = 0$ for even n (see Fig. 2). Note that the addition of the term (6) lifts all the zeros of the function (4) to a slightly positive value – except one. Numerical results show that the minimal cv is obtained when the first zero remains unlifted. See Fig. 2.

The reason the cv value changes little when going from odd n to even $n + 1$ is that the gain by the change $f(0) = 0$ is smaller in scale than what is gained in concentration by the introduction of a new $\cos^2$ term.

To summarize, we look for an optimal $f(t)$ in the form (5) with the additional properties

– $f(0) = 0$;
– $f(x) = 0$ and $f'(x) = 0$ for some $x > 0$.

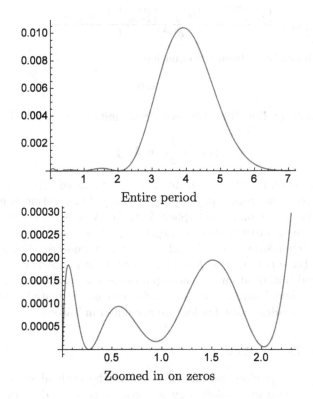

Fig. 2. Optimal f for $n = 8$

3.3 Optimization for Even n

The difficulty in optimizing $f(t)$ in (5) for an even n is that nonnegativity of $f(t)$ for $t \geq 0$ is not guaranteed and is difficult to check in general. To compute $f(t)$ with parameters with minimal cv, we reparametrize $f(t)$ in the following way.

We eliminate a_2 by solving $f(0) = 0$; thus

$$a_2 = -a_3 - \prod_{i=0}^{(n-2)/2} \cos^2(\phi_i).$$

Next we replace the variable a_3 by x where x is the zero of f; that is, solve the equation

$$f(x) = 0$$

for a_3. The solution is explicit for a given n:

$$a_3 = \frac{\prod_{i=0}^{(n-2)/2} \cos^2(\phi_i) - \prod_{i=0}^{(n-2)/2} \cos^2(\omega x - \phi_i)}{e^{x(\lambda_2+1)} - 1}.$$

Next we eliminate ϕ_1 from the equation

$$f'(x) = 0 \tag{7}$$

in the following way. Rewriting the terms containing ϕ_1 using the formula

$$\cos^2(\phi_1) = \frac{1}{2}(\cos(2\phi_1) + 1),$$

(7) leads to an equation that is linear in $\cos(2\phi_1)$ and $\cos(2(\omega x - \phi_1))$, which in turn leads to a quadratic equation for $\cos(2\phi_1)$. The solution is explicit but prohibitively long – we omit the explicit formula. We note that only one of the solutions corresponds to a proper nonnegative $f(t)$ in (5).

After these transformations, f (and thus cv) is parametrized by $x, \omega, \phi_2, \ldots, \phi_{n/2-1}$. With these parameters, f is guaranteed to be nonnegative.

A numerical optimization for these parameters is now feasible; after optimization, the original parameters (a_2, a_3, ϕ_1) can be calculated explicitly. The results of the numerical optimization are presented in Table 3.

3.4 Rate of Decay of cv

We numerically determine the rate of decay of the optimal value of cv as a function of n (the optimal value of cv for order n is denoted by $\mathrm{cv}(n)$). Based on the cv values in Table 1 and due to the different behavior of $\mathrm{cv}(n)$ for odd and even values of n as discussed in Sect. 3.2, we only consider odd values of n. Table 1 suggests a polynomial decay in n and this trend is tested by plotting $\mathrm{cv}(n)$ against n with log-log scaling in Fig. 3. The figure shows that the decay

Table 3. Optimal parameter values for even n

Poles	λ_2	a_2	a_3	ω	$\phi_1, \phi_2, \phi_3, \ldots$
	Parameters				
4	1.94907	0.224603	-0.589603	0.519765	2.2195
6	-8.34112	0.00074	-0.01551	0.477957	1.78917; 2.1665
8	-13.917	4.50119×10^{-5}	-1.93645×10^{-3}	0.44404	1.70917; 1.98567; 2.46912
10	-20.01365	4.53552×10^{-6}	-3.15042×10^{-4}	0.419619	1.66911; 1.88536; 2.25959; 2.72108
12	-26.64365	5.89469×10^{-7}	-5.89972×10^{-5}	0.40078	1.64553; 1.82172; 2.12538; 2.5004; 2.9258
14	-33.78365	8.89442×10^{-8}	-1.20051×10^{-5}	0.385688	1.63023; 1.77798; 2.03236; 2.34773; 2.704288; 3.09330

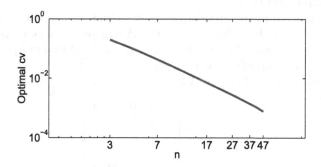

Fig. 3. Decay of $\mathrm{cv}(n)$ on a log-log scale

is very close to linear on log-log scale, with gradient -2.03. Least square fitting gives the approximation

$$\mathrm{cv}(n) \approx \frac{2.175}{n^{2.03}},$$

but for a simple yet relatively accurate formula, one can also use

$$\mathrm{cv}(n) \sim \frac{2}{n^2}.$$

4 Conclusion

Following [10], we have made further numerical investigations for concentrated higher order ME distributions. We have obtained numerical results for a certain subclass conjectured to contain the ME distribution with minimal cv of ME distributions of odd order up to order 47 and for even order up to order 14. We also expanded and improved the optimization methods. We also provided the parameters of the extreme distributions, and compared the (numerical and abstract) properties of the optimal ME distributions for odd and even order.

As main conclusions we found that

- the minimal cv of even order n ME distributions gets to be very close to the one of order $n - 1$ as n is increasing, and
- the minimal cv of odd order n ME distribution is close to $\frac{2}{n^2}$.

A Various Forms of Matrix Exponential Functions

This section is dedicated to show the equivalence of the various forms of matrix exponential functions throughout the paper. Specifically, we show the equivalence of the forms (2) and (3) and also show how (4) (and also (5)) can be brought to a form consistent with (2) and (3).

As mentioned in Sect. 2, (2) can be converted directly into (3) using the Jordan-decomposition of A.

From a pdf given in the form (3), one can reconstruct a matrix-vector representation in (2) in the following manner: A will be in block-diagonal form, with each block corresponding to either a single real eigenvalue λ_j or a pair of complex eigenvalues $\lambda_j, \lambda_{j+1} = a \pm \mathcal{I}b$, where $\mathcal{I} = \sqrt{-1}$.

If λ_j is real, then the block in A is

$$[\lambda_j] \quad \text{and} \quad \begin{bmatrix} \lambda_j & 1 & 0 & \dots & 0 \\ 0 & \lambda_j & 1 & \dots & 0 \\ \vdots & & \ddots & & \\ & & & \lambda_j & 1 \\ 0 & & \dots & 0 & \lambda_j \end{bmatrix}$$

for multiplicity 1 and $N_j > 1$, respectively.

For a complex pair of eigenvalues $\lambda_j, \lambda_{j+1} = a \pm \mathcal{I}b$, the block is

$$\begin{bmatrix} a & b \\ -b & a \end{bmatrix} \quad \text{or} \quad \begin{bmatrix} a & b & 1 & 0 & & & & 0 \\ -b & a & 0 & 1 & 0 & & \dots & 0 \\ 0 & 0 & a & b & 1 & 0 & & 0 \\ 0 & 0 & -b & a & 0 & 1 & 0 & 0 \\ 0 & & & & & & & \\ \vdots & & & & \ddots & & & \vdots \\ & & & & & & a & b \\ 0 & & & \dots & & & -b & a \end{bmatrix}$$

for multiplicity 1 or $N_j > 1$ respectively (the matrix on the right is size $2N_j \times 2N_j$). Once A is constructed, α can be obtained by solving a system of linear equations.

Finally, (4) can be represented in a form consistent with (3); the eigenvalues are $-1, (-1 \pm 2\mathcal{I}\omega), \ldots, (-1 \pm (n-1)\mathcal{I}\omega)$. We demonstrate this for $n = 5$; for higher odd values of n, it follows a similar structure, albeit with more terms:

$$
\begin{aligned}
f(t) =& e^{-t} \cos^2(\omega t - \phi_1) \cos^2(\omega t - \phi_2) \\
=& \frac{1}{8} e^{-t} \big(2 + \cos(2\phi_1 - 2\phi_2)\big) + \frac{1}{8} e^{-t} \big(\cos(2\phi_1) \cos(2t\omega) \\
& + \sin(2\phi_1) \sin(2t\omega) + \cos(2\phi_2) \cos(2t\omega) + \sin(2\phi_2) \sin(2t\omega)\big) \\
& + \frac{1}{8} e^{-t} \big(\cos(2\phi_1 + 2\phi_2) \cos(4t\omega) + \sin(2\phi_1 + 2\phi_2) \sin(4t\omega)\big) \\
=& \frac{1}{8} e^{-t} (2 + \cos(2\phi_1 - 2\phi_2)) + \\
& + \frac{1}{8} (e^{2i\phi_1} + e^{2i\phi_2}) e^{(-1-2i\omega)t} + \frac{1}{8} (e^{-2i\phi_1} - e^{2i\phi_2}) e^{(-1+2i\omega)t} + \\
& \frac{1}{16} e^{2i(\phi_1 + \phi_2)} e^{2t(-1+4i\omega)} + \frac{1}{16} e^{-2i(\phi_1 + \phi_2)} e^{2t(-1-4i\omega)}.
\end{aligned}
$$

Representing (5) in a form consistent with (3) is essentially the same, just with one extra real eigenvalue.

B Optimal Parameter Values

Table 4. Optimal parameter values for odd values $19 \leq n \leq 47$

19	0.353490	0.29829; 1.59508; 1.69721; 1.86521; 2.07671;2.31637; 2.57570; 2.85029; 3.13827
21	0.345640	0.10490; 0.38461; 1.59232; 1.68321; 1.83365; 2.02382; 2.23966; 2.47319; 2.72005; 2.97806
23	0.338715	0.19898; 0.45955; 1.59006; 1.67176; 1.80774; 1.98031; 2.17653; 2.38888; 2.61318; 2.84711; 3.08961
25	0.332545	0.04609; 0.281467; 0.52523; 1.58820; 1.66224; 1.78614; 1.94394; 2.12371; 2.31837; 2.52389; 2.73798; 2.95940
27	0.327002	0.13289; 0.35436; 0.58327; 1.58663; 1.65422; 1.767866; 1.91313; 2.07891; 2.25855; 2.44819; 2.64559; 2.84945; 3.05913
29	0.31348	0.03935; 0.24334; 0.45271; 1.14455; 1.58818; 1.65855; 1.77085; 1.91037; 2.06735; 2.23615; 2.41354; 2.59761; 2.78723; 2.98175
31	0.308959	0.11601; 0.30925; 0.50731; 1.16023; 1.58673; 1.65148; 1.75543; 1.88509; 2.03128; 2.18865; 2.35408; 2.52570; 2.70239; 2.88346; 3.06855

(continued)

Table 4. (*continued*)

33	0.304832	0.005513; 0.18531; 0.36880; 0.55668; 1.17491; 1.58548; 1.64538; 1.74205; 1.86308; 1.99983; 2.14712; 2.30215; 2.46291; 2.62834; 2.79775; 2.97073
35	0.301040	0.07689; 0.24812; 0.42288; 0.60154; 1.18864; 1.58439; 1.64006; 1.73034; 1.84377; 1.97218; 2.11069; 2.25642; 2.40761; 2.56315; 2.72235; 2.88477; 3.05017
37	0.297540	0.14201; 0.30558; 0.47218; 0.64248; 1.20148; 1.58344; 1.63540; 1.72002; 1.82669; 1.94769; 2.07833; 2.21585; 2.35853; 2.50531; 2.65547; 2.80859; 2.96438; 3.12272;
39	0.294289	0.04762; 0.20164; 0.35809; 0.51732; 0.67999; 1.21350; 1.58260; 1.63127; 1.71086; 1.81149; 1.92585; 2.04945; 2.17962; 2.31470; 2.45364; 2.59576; 2.74061; 2.88789; 3.03744
41	0.291265	0.10875; 0.25643; 0.40632; 0.55879; 0.71449; 1.62759; 1.22476; 1.58186; 1.70267; 1.79788; 1.90627; 2.02352; 2.14707; 2.27531; 2.40722; 2.54211; 2.67955; 2.81924; 2.96098; 3.10469
43	0.286709	0.017601; 0.15905; 0.29899; 0.44498; 0.59176 0.74070; 1.23092; 1.58084; 1.62373; 1.69468; 1.78474; 1.88730; 1.99909; 2.12208; 2.29038; 2.30290; 2.47864; 2.61584 2.74717; 2.88339; 3.02186
45	0.276478	0.08675; 0.22262; 0.35787; 0.498054; 0.63799; 1.08335; 1.25082; 1.58080; 1.62305; 1.69202; 1.77581; 1.87951; 1.98293; 2.09473; 2.20973; 2.32692; 2.45504; 2.57721; 2.70603; 2.83471; 2.96499; 3.09616
47	0.283839	0.11500; 0.23946; 0.37625; 0.51191; 0.64863; 1.07858; 1.23819; 1.57945; 1.61595; 1.65715; 1.76319; 1.85467; 1.95591; 2.06011; 2.16775; 2.29640; 2.37101; 2.56629; 2.62701; 2.70473; 2.92734; 2.94086; 3.11827

References

1. Code for numerical optimization in Matlab. http://webspn.hit.bme.hu/~illes/mincvnum.zip. Accessed 27 July 2016
2. Bean, N.G., Nielsen, B.F.: Quasi-birth-and-death processes with rational arrival process components. Stoch. Models **26**(3), 309–334 (2010)
3. Commault, C., Mocanu, S.: Phase-type distributions and representations: some open problems for system theory. Int. J. Control **76**(6), 566–580 (2003)
4. David, A., Larry, S.: The least variable phase type distribution is Erlang. Stoch. Models **3**(3), 467–473 (1987)
5. Éltető, T., Rácz, S., Telek, M.: Minimal coefficient of variation of matrix exponential distributions. In: 2nd Madrid Conference on Queueing Theory, Madrid, Spain, July 2006. Abstract
6. Horváth, I., Telek, M.: A constructive proof of the phase-type characterization theorem. Stoch. Models **31**(2), 316–350 (2015)
7. Maier, R.S.: The algebraic construction of phase-type distributions. Commun. Stat. Stoch. Models **7**(4), 573–602 (1991)

8. Mocanu, S., Commault, C.: Sparse representations of phase-type distributions. Commun. Stat. Stoch. Models **15**(4), 759–778 (1999)
9. Colm Art O'Cinneide: Characterization of phase-type distributions. Commun. Stat. Stoch. Models **6**(1), 1–57 (1990)
10. Horvath, A., Buchholz, P., Telek, M.: Stochastic Petri nets with low variation matrix exponentially distributed firing time. Int. J. Perform. Eng. **7**, 441–454 (2011)
11. Rechenberg, I.: Evolutionstrategie: Optimierung technisher Systeme nach Prinzipien des biologischen Evolution. Frommann-Hollboog Verlag, Stuttgart (1973)
12. van de Liefvoort, A.: The moment problem for continuous distributions. Technical report, WP-CM-02, University of Missouri, Kansas City (1990)

A Stochastic Model-Based Approach to Online Event Prediction and Response Scheduling

Marco Biagi[(✉)], Laura Carnevali, Marco Paolieri, Fulvio Patara, and Enrico Vicario

Department of Information Engineering, University of Florence, Florence, Italy
marco.biagi@unifi.it

Abstract. In a variety of contexts, time-stamped and typed event logs enable the construction of a stochastic model capturing the sequencing and timing of observable discrete events. This model can serve various objectives including: diagnosis of the current state; prediction of its evolution over time; scheduling of response actions. We propose a technique that supports online scheduling of actions based on a prediction of the model state evolution: the model is derived automatically by customizing the general structure of a semi-Markov process so as to fit the statistics of observed logs; the prediction is updated whenever any observable event changes the current state estimation; the (continuous) time point of the next scheduled action is decided according to policies based on the estimated distribution of the time to given critical states. Experimental results are reported to characterize the applicability of the approach with respect to general properties of the statistics of observable events and with respect to a specific reference dataset from the context of Ambient Assisted Living.

1 Introduction

The prediction of future system events from past ones is a challenging and general problem that finds several applications (e.g., *autonomic computing* [1,9], *models at runtime* [2]), especially in online settings where predictions can be updated several times after observing new data [15].

Model-based approaches have also been proposed in other application areas such as Ambient Assisted Living (AAL). An important goal in AAL is to recognize human activities from smart home sensor data. Common activities of interest are Activities of Daily Living (ADLs) such as "bathing", "sleeping", "dinner"; home appliances and furniture can generate sensor events indicating, for example, the use of a faucet, the opening of a door, or the use of a light switch. The problem of assessing the current human activity, also known as *diagnosis*, has been investigated in [4,16] through the use of hidden stochastic models.

In this work, we are interested in the problem of *event prediction* and *response scheduling*: given a sequence of past events, each with a type and a timestamp, our goal is to select a time point for the activation of a response action. A response action can, for example, replace a hardware component to prevent

© Springer International Publishing AG 2016
D. Fiems et al. (Eds.): EPEW 2016, LNCS 9951, pp. 32–47, 2016.
DOI: 10.1007/978-3-319-46433-6_3

failures; in AAL, the response action can be a reminder about the intake of some drug before the beginning of a meal, or surveillance escalation when the user is going to use some dangerous device.

To this end, we model the system under analysis as a Semi-Markov Process (SMP) where states represent activities or *idle times* between activities. For each state, the sojourn time distribution and transition probabilities are estimated from a supervised dataset of event logs. Different parametric models are compared for sojourn time distributions, in order to evaluate the reduction in the prediction error due to more accurate models; in particular, we consider fitting the mean of the sojourn time with an exponentially distributed (EXP) or Erlang random variable, or fitting its mean and variance using a shifted EXP random variable, or sums and mixtures of EXP random variables [17]. After fitting the model parameters from a training dataset, we use a test set of events to assess the prediction error of scheduling policies for the response action. After each event of the test set, the current state and elapsed sojourn time (i.e., a diagnosis) are used as initial conditions to compute first-passage probabilities for critical states (e.g., dinner). Scheduling policies analyze this transient information to select an actuation time for the response action.

Related work. In [4], we proposed a model-driven online approach to the problem of diagnosis: a model of feasible behaviors, enhanced with stochastic parameters derived from a dataset [13], was used to evaluate a measure of likelihood of the current ongoing activity, which is updated after each event. A different approach to diagnosis is considered in [16], where hidden semi-Markov processes are used to detect the current user activity from recent events; in [15], hidden semi-Markov processes are used to predict and prevent imminent failures; in [7], Markov-modulated Poisson processes are proposed to detect anomalous patterns. In [12], a short-term prediction problem is considered: the next state is predicted by combining different information sources through Dempster-Shafer theory. Our work takes a different approach: instead of matching the sequence of recent events to determine the current or next state, we use a state diagnosis as the initial condition to analyze the transient evolution of a semi-Markov process and derive first-passage probabilities of critical states using the ORIS Tool [3,6]. This provides fine-grained information for policies that schedule response actions.

Organization. In Sect. 2, we formulate the prediction and scheduling problem; in Sect. 3, we define the structure, fitting, and analysis technique for the SMP model adopted by our solution. In Sect. 4, we evaluate the prediction error of the approach, both on synthetic datasets and on a real-world dataset of ADLs. Finally, we draw our conclusions in Sect. 5.

2 Problem Formulation

We consider an online setting where the system receives a stream of *events* (e.g., sensor readings), each with a type and a timestamp; after receiving an event, the system has the opportunity to request or update the scheduling of a *response action* (e.g., surveillance escalation or a user reminder), to be activated after a delay if no other event is received. The goal is to intercept a class of *target activities* (e.g., a security attack, a hardware fault, the dinner activity or other ADLs) with the response action. The response action is active for a fixed *response duration* $t_d \geq 0$: target activities are successfully handled by the system if they start while the response action is active. In order to cope with real-world applications, we also introduce a fixed *response actuation time* $t_w \geq 0$: the response action must be issued by the system t_w time units before its activation time; after being issued, response actions cannot be canceled. For example if a hardware failure is predicted in the future, a maintenance operation must be scheduled t_w time units before it happens, so as to prevent the failure of the system.

In order to learn the relationship between input events and target activities, we are given a supervised dataset $\mathcal{D}$ including: *(i)* a recorded sequence of events $e_1, e_2, \ldots, e_m$ where each event $e_i = \langle \varepsilon_i, t_i \rangle$ for $i = 1, \ldots, m$ includes an event type ε_i from a finite set $\mathcal{E} = \{\mathtt{evt}_1, \ldots, \mathtt{evt}_M\}$ and a timestamp $t_i \geq 0$, with $t_1 \leq \cdots \leq t_m$; *(ii)* the sequence of activities $a_1, a_2, \ldots, a_n$ performed during the generation of the events; each activity $a_i = \langle \alpha_i, \tau_i, \delta_i \rangle$ for $i = 1, \ldots, n$ includes an activity type α_i from a finite set $\mathcal{A} = \{\mathtt{act}_1, \ldots, \mathtt{act}_N\}$, a start time $\tau_i \geq 0$ (such that $\tau_1 \leq \cdots \leq \tau_n$) and an activity duration $\delta_i \geq 0$. As in [4,16], the underlying assumption of this problem is that distinct activities take place in the system and a statistical characterization is derived taking into account: *(i)* the duration of each activity; *(ii)* the time between subsequent events; *(iii)* the types of event occurred within each activity; *(iv)* the transition probabilities among activities. The sequence of activities thus provides a "ground truth" to learn the relationship between events and activities, and the likelihood of different activity sequences.

In order to evaluate the online performance of a scheduling system $f_{\mathcal{D}}$ trained from $\mathcal{D}$, we define the metrics of *precision* and *recall* for a given test sequence of events $e_1, e_2, \ldots, e_h$, an actuation time $t_w \geq 0$, and a response duration $t_d \geq 0$. Let $\tilde{t}$ denote the time point at which the system is currently scheduled to issue the activation of the response action, which will be active in the time window $[\tilde{t} + t_w, \tilde{t} + t_w + t_d]$. Initially, no response activation is scheduled, i.e., $\tilde{t} = \infty$; for each event $e_i = \langle \varepsilon_i, t_i \rangle$ in the sequence $e_1, e_2, \ldots, e_h$: *(i)* if $\tilde{t} \leq t_i$, the activation of a response action is issued at time $\tilde{t}$; *(ii)* the time point $\tilde{t}$ is updated by $f_{\mathcal{D}}$ according to the new information provided by event e_i, i.e., $\tilde{t} \leftarrow f_{\mathcal{D}}(e_1, e_2, \ldots, e_i; t_w, t_d)$. Let $I_1, I_2, \ldots, I_p$ denote the time windows of executed response actions and let $\overline{\tau}_1, \overline{\tau}_2, \ldots, \overline{\tau}_q$ denote the start times of target activities in the ground truth. We say that the system has produced: *(i)* a *true positive*, if an activation window contained the start time of (at least) a target activity, i.e., $\mathrm{TP} = |\{i \leq p \mid \exists j \leq q \text{ such that } \overline{\tau}_j \in I_i\}|$; *(ii)* a *false positive*, if an activation window did not contain any target activity, i.e., $\mathrm{FP} = |\{i \leq p \mid \overline{\tau}_j \notin I_i \, \forall j \leq q\}|$ *(iii)* a *false negative*, if a target activity was not contained in any activation window, i.e., $\mathrm{FN} = |\{j \leq q \mid \overline{\tau}_j \notin I_i \, \forall i \leq p\}|$ Then, we define (as usual in information retrieval) *precision* $= \mathrm{TP}/(\mathrm{TP} + \mathrm{FP})$ and *recall* $= \mathrm{TP}/(\mathrm{TP} + \mathrm{FN})$.

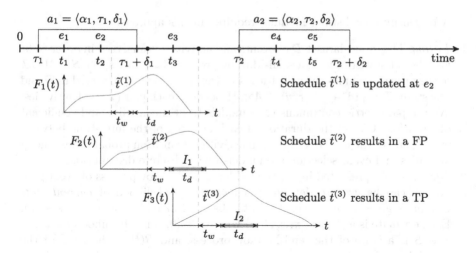

Fig. 1. A sample execution of the online scheduling system: α_2 is the target activity; the start of response actions is issued at $\tilde{t}^{(2)}$ and $\tilde{t}^{(3)}$, producing the activity windows I_1 and I_2 (a false positive and a true positive, respectively).

Figure 1 illustrates the mechanism of online event prediction and response scheduling on the sequences of events e_1, e_2, e_3, e_4, e_5 and activities a_1, a_2. The type $\alpha_2 \in \mathcal{A}$ of $a_2 = \langle \alpha_2, \tau_2, \delta_2 \rangle$ is the target activity: after event e_1, the system uses a metric $F_1(t)$ to select the best activation window of duration t_d for the response action; the activation command is first scheduled at time $\tilde{t}^{(1)}$. Since event e_2 is received at time $t_2 < \tilde{t}^{(1)}$, a new metric $F_2(t)$ and schedule $\tilde{t}^{(2)}$ is computed; given that no event is received before $\tilde{t}^{(2)}$, the response action is issued, resulting in the activation window $I_1 = [\tilde{t}^{(2)} + t_w, \tilde{t}^{(2)} + t_w + t_d]$. This response is a "false positive", because it does not intercept the start of any target activity α_2. At event e_3, the metric $F_3(t)$ and schedule $\tilde{t}^{(3)}$ are computed. As $t_4 > \tilde{t}^{(3)}$ (i.e., event e_4 is received after $\tilde{t}^{(3)}$), the response action is activated during the time interval $I_2 = [\tilde{t}^{(3)} + t_w, \tilde{t}^{(3)} + t_w + t_d]$. The interval I_2 contains the start time τ_2 of a target activity, resulting in a "true positive" response.

Our formulation is similar to that of [14] for the problem of online failure prediction, as it considers both an actuation time t_w and a response duration t_d, which are constant parameters of the specific application. Nonetheless, in contrast to [14], it requires that the system selects the time point $\tilde{t}$ that corresponds to the best schedule for the activation window $[\tilde{t} + t_w, \tilde{t} + t_w + t_d]$.

3   A Model-Based Solution

3.1   System Architecture

We propose a solution based on a stochastic model capturing the statistics of events and activities observed in the training dataset. The model allows us to estimate the current state and predict the occurrence of target activities.

We identify four distinct modules required in this approach.

1. *Fitting.* The input dataset $\mathcal{D}$ of events and activities is analyzed in order to fit its relevant statistics with a semi-Markov process. The state space $S = \mathcal{A} \cup \mathcal{I}$ of the process includes the activities $\mathcal{A} = \{\mathtt{act}_1, \ldots, \mathtt{act}_N\}$ observed in $\mathcal{D}$ and idle states $\mathcal{I} = \{\mathtt{idle}_{xy} \mid (x, y) \in \mathcal{A} \times \mathcal{A}\}$ between each pair (x, y) of activities. We use parametric continuous-time models to fit mean value and coefficient of variation (CV) of the duration of each activity, of the inter-time between events, and of the idle times between activities; transition probabilities among activities and event selection are modeled with discrete distributions.

2. *Diagnosis.* As presented in [4], the hidden semi-Markov process of event generation can be used to determine a measure of likelihood of *current state estimates* given a sequence of events observed in real-time by the system. Each estimate is a triplet $\langle \pi, x, R(t) \rangle$ where $\pi \in [0, 1]$ is a likelihood measure, $x \in S$ is a state of the semi-Markov process, and $R(t)$ is the PDF of the remaining sojourn time in x (estimated numerically on a grid of time points).

3. *Prediction.* From each estimate of the current state, we analyze the transient evolution of the semi-Markov process, computing first-passage transient probabilities for a set of target activities. First-passage probabilities computed for each current state estimate are then combined according to their likelihoods, in order to obtain the expected probability $F(t)$ of reaching some critical state within time t.

4. *Scheduling.* The first-passage probabilities $F(t)$ are analyzed according to a *policy* that selects a time point where the activation of the response action is scheduled. As input parameters of the problem, the policy considers the response actuation time $t_w \geq 0$, required to activate the response action, and the response duration $t_d \geq 0$.

Figure 2 presents a data flow diagram illustrating the components of the system and the information exchanged as input and output. By decomposing the system architecture, we are able to isolate the responsibilities of the components

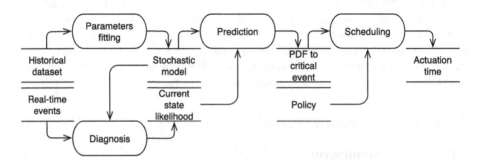

Fig. 2. Data flow diagram of the proposed solution: events are processed in real-time to assess the current state and evaluate first-passage probabilities of a set of critical activities; response actions are then scheduled based on a given policy.

and evaluate their individual performance. While [4] focused on the diagnosis of the current state, in this work we study the prediction and scheduling components. In Sect. 4, we will use an exact state diagnosis (from the ground truth of recorded activities) as input for the prediction phase; in so doing, we will decouple the performance of prediction and scheduling from that of state diagnosis.

3.2 Model Definition

We model the system under analysis as a semi-Markov process where states represent activities or *idle times* between activities.

Definition 1 (Markov Renewal Sequence). *A sequence of random variables* $\{(X_n, T_n),\ n \in \mathbb{N}\}$ *such that, for all* $n \in \mathbb{N}$, X_n *takes values in a finite set* S, T_n *takes values in* $\mathbb{R}_{\geqslant 0}$, *and* $0 = T_0 \leq T_1 \leq T_2 \leq \cdots \leq T_n$, *is called* Markov renewal sequence *with state space* S *and kernel* $G_{ij}(t)$ *if and only if*

$$P\{X_{n+1} = j,\ T_{n+1} - T_n \leq t \mid X_n = i, X_{n-1}, \ldots, X_1, X_0, T_n, \ldots, T_1, T_0\}$$
$$= P\{X_1 = j,\ T_1 \leq t \mid X_0 = i\} := G_{ij}(t)$$

for all $n \in \mathbb{N}$, $i, j \in S$ *and* $t \in \mathbb{R}_{\geqslant 0}$.

The Markov renewal sequence is thus time-homogeneous and memoryless: given the current state $i \in S$, the kernel $G_{ij}(t)$ defines the joint distribution of the next *renewal time* T_1 and *regeneration state* $X_1 \in S$. From a Markov renewal sequence, a semi-Markov process can be constructed as follows.

Definition 2 (Semi-Markov Process). *Let* $\{(X_n, T_n),\ n \in \mathbb{N}\}$ *be a Markov renewal sequence with state space* S; *we define* semi-Markov process *the process* $\{X(t),\ t \geq 0\}$ *such that* $X(t) = X_n$ *for* $t \in [T_n, T_{n+1})$ *and all* $n \in \mathbb{N}$.

Given a training dataset of N activities $\mathcal{A} = \{\mathsf{act}_1, \ldots, \mathsf{act}_N\}$, we construct a semi-Markov process with state space $S = \mathcal{A} \cup \mathcal{I}$, where $\mathcal{I} = \{\mathsf{idle}_{xy} \mid (x, y) \in \mathcal{A} \times \mathcal{A}\}$ is the set of "idle states" between each pair (x, y) of activities. The state of the system evolves as follows: *(i)* the system randomly selects a sojourn time in the current activity $\mathsf{act}_x \in \mathcal{A}$ and the successive idle state $\mathsf{idle}_{xy} \in \mathcal{I}$ for some $y \in \mathcal{A}$; *(ii)* a sojourn time is selected for the state idle_{xy}; *(iii)* from idle_{xy}, the model moves to the state corresponding to $\mathsf{act}_y \in \mathcal{A}$. This structure of the semi-Markov process is reflected in the definition of its kernel

$$G_{ij}(t) := \begin{cases} H_x(t)\, p_{xy} & \text{if } i = \mathsf{act}_x \in \mathcal{A} \text{ and } j = \mathsf{idle}_{xy} \in \mathcal{I} \text{ for some } y \in \mathcal{A}, \\ D_{xy}(t) & \text{if } i = \mathsf{idle}_{xy} \in \mathcal{I} \text{ and } j = \mathsf{act}_y \in \mathcal{A}, \\ 0 & \text{otherwise,} \end{cases}$$

where: $H_x(t)$ is the distribution of the sojourn time in act_x; p_{xy} is the transition probability from activity act_x to activity act_y, with $\sum_{y \in \mathcal{A}} p_{xy} = 1$ for all $x \in \mathcal{A}$; $D_{xy}(t)$ is the sojourn time distribution in the idle state idle_{xy} between act_x and act_y. Figure 3 illustrates the structure of this semi-Markov model. Note that,

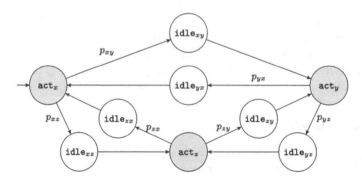

Fig. 3. Structure of the semi-Markov process for the activities $\mathcal{A} = \{\text{act}_x, \text{act}_y, \text{act}_z\}$.

in our definition, the transition probabilities p_{xy} do not depend on the sojourn time in the current activity. Nonetheless, the definition of the semi-Markov kernel $G_{ij}(t)$ and the prediction and scheduling approach described in the following can model this dependency; we adopt time-homogeneous transition probabilities p_{xy} due to the limited amount of data available for fitting in real-world applications.

3.3   Parameter Fitting from a Training Dataset

The stochastic parameters of the model are automatically derived from the statistics of activities observed in the dataset. Specifically, we are interested in characterizing three aspects: (1) the sojourn time distribution $H_x(t)$ in each activity $\text{act}_x \in \mathcal{A}$; (2) the sojourn time distribution $D_{xy}(t)$ in each idle state $\text{idle}_{xy} \in \mathcal{I}$; and, (3) the transition probability p_{xy} from each activity act_x to any other reachable activity act_y.

Let $a_i = \langle \alpha_i, \tau_i, \delta_i \rangle$ for $i = 1, \ldots, n$ be the sequence of activities in the training dataset, each including an activity type $\alpha_i \in \mathcal{A}$, a start time $\tau_i \geq 0$, and an activity duration $\delta_i \geq 0$.

For each activity $\text{act}_x \in \mathcal{A}$, we consider its observed durations $\delta(\text{act}_x) = \{\delta_i \mid 1 \leq i \leq n \text{ and } \alpha_i = \text{act}_x\}$ and estimate the sojourn time distribution $H_x(t)$ using one of three different strategies:

1. *Exp strategy.* Fitting the mean value μ of $\delta(\text{act}_x)$ with an exponentially distributed random variable with rate $\lambda = 1/\mu$.
2. *Erlang strategy.* Fitting the mean value μ of $\delta(\text{act}_x)$ with an Erlang random variable with shape parameter $k = 2$ and rate $\lambda = k/\mu$.
3. *Whitt strategy.* Fitting the mean value μ and coefficient of variation CV of $\delta(\text{act}_x)$ with the approach presented in [17], which requires: *(i)* if $\text{CV} \leq 1/\sqrt{2}$, a shifted exponential random variable with PDF $f(t) = \lambda e^{-\lambda(t-d)}$ over $[d, \infty)$, where $\lambda = 1/(\mu \, \text{CV})$ and $d = \mu(1 - \text{CV})$; *(ii)* if $1/\sqrt{2} < \text{CV} < 1$, a hypo-exponential random variable (sum of two exponential random variables) with PDF $f(t) = \lambda_1 \lambda_2 (e^{-\lambda_2 t} - e^{-\lambda_1 t})/(\lambda_1 - \lambda_2)$ with $\lambda_i = 1/[(\mu/2)(1 \pm \sqrt{2 \, \text{CV}^2 - 1})]$ for $i = 1, 2$; *(iii)* if $\text{CV} > 1$, a hyper-exponential

random variable with PDF $f(t) = p_1\lambda_1e^{-\lambda_1 t} + p_2\lambda_2e^{-\lambda_2 t}$, where $p_i = [1 \pm \sqrt{\frac{CV^2-1}{CV^2+1}}]/2$ and $\lambda_i = 2p_i/\mu$ for $i = 1, 2$.

We adopt the same strategies for the estimation of the sojourn time distribution $D_{xy}(t)$ of each idle state $\text{idle}_{xy} \in \mathcal{I}$; in this case, the mean value and coefficient of variation of the sojourn in idle_{xy} are computed from the observed durations $\{\tau_{i+1} - (\tau_i + \delta_i) \mid 1 \leq i < n, \alpha_i = \text{act}_x \text{ and } \alpha_{i+1} = \text{act}_y\}$.

Finally, the transition probability p_{xy} that activity act_y will follow act_x is estimated as

$$p_{xy} = \frac{|\{1 \leq i < n \mid \alpha_i = \text{act}_x \text{ and } \alpha_{i+1} = \text{act}_y\}|}{|\{1 \leq i < n \mid \alpha_i = \text{act}_x\}|}$$

for each pair of activities $(\text{act}_x, \text{act}_y)$ in the training dataset.

3.4   Prediction

After receiving an event, a new set of state estimates $\langle \pi_u, x_u, R_u(t)\rangle$, $u = 1, \ldots, h$ is produced by the diagnosis component, each with a likelihood measure $\pi_u \in [0, 1]$, a candidate state $x_u \in S$, and a PDF $R_u(t)$ of the remaining sojourn time in x_u. The prediction component uses this information on the current state of the system in order to compute first-passage probabilities for a set of target activities using the semi-Markov model.

Let $S = \mathcal{A} \cup \mathcal{I}$ be the set of states of the SMP (i.e., activities and idle states) and let $G_{ij}(t)$ be its kernel: for all $i, j \in S$ and $t \geq 0$, $G_{ij}(t)$ gives the probability that the next state j is reached from i in a time lower or equal to t. We compute first-passage probabilities of a set of target activities $A \subset \mathcal{A}$ by solving a system of Markov renewal equations [11]

$$P_{ij}(t) = \left(1 - \sum_{j \in S} G_{ij}^A(t)\right)\delta_{ij} + \sum_{k \in S} \int_0^t dG_{ik}^A(x) P_{kj}(t - x)$$

for all $i, j \in S$ and $0 \leq t \leq t_{max}$, where t_{max} is the time bound of the analysis, $\delta_{ij} = 1$ for $i = j$ and $\delta_{ij} = 0$ otherwise, and

$$G_{ij}^A(t) = \begin{cases} 0 & \text{if } i \in A, \\ G_{ij}(t) & \text{if } i \notin A, \end{cases}$$

effectively making any state in A an *absorbing state*. A numerical solution based on the trapezoidal rule [6] requires $O((\frac{t_{max}}{\Delta})^2)$ multiplications of $|S| \times |S|$ matrices, where Δ is the step size used in the discretization. The first-passage probabilities of states in A are then given, for each initial state $i \in S$, by $F_i(t) = \sum_{j \in A} P_{ij}(t)$.

We numerically compute $F_i(t)$ only once for a fixed time bound t_{max} and each $i \in S$. After a new event, the updated diagnosis $\langle \pi_u, x_u, R_u(t)\rangle$ for $u = 1, \ldots, h$ is used to compute

$$F(t) = \sum_{u=1}^h \pi_u \int_0^{t_{max}} R_u(x) \left(\sum_{j \in S} G_{x_u, j}(\infty) F_j(t - x)\right) dx$$

which accounts for each state x_u according to its likelihood π_u, remaining sojourn time PDF $R_u(x)$, transition probabilities $G_{x_u,j}(\infty)$ to the next state j, and first-passage probability $F_j(t)$ from j to a state in A.

3.5 Response Scheduling

Given $F(t)$ and a policy, the scheduling component (Fig. 2) selects a response actuation time $\tilde{t}$. In this work, we consider a *maximum probability interval* policy. The policy considers a discrete grid of points $\mathcal{X}$ equispaced in $[t_w, t_{max} - t_d]$: for each $x \in \mathcal{X}$, the probability

$$v(x) = F(x + t_d) - F(x)$$

of reaching a target activity in the interval $[x, x + t_d]$ is evaluated. Let $x^* = \text{argmax}_{x \in \mathcal{X}} v(x)$. If $v(x^*) > \epsilon t_d$, then the activation of the response action is scheduled at time $\tilde{t} = x^* - t_w$; otherwise, no response action is scheduled.

4   Experimental Evaluation

An experimentation was carried out to evaluate the system performance in terms of precision and recall metrics. We experimented both with synthetic datasets constructed so as to make evident how the type of the distribution can affect the prediction and scheduling performance, and with a real dataset [16] so as to evaluate the applicability of the proposed approach in a problem of Activity Recognition (AR) from the context of Ambient Assisted Living (AAL).

In both cases, an exact state diagnosis was assumed as input for the prediction phase, so as to focus on the evaluation of performance achieved by predictor and scheduler components.

Each dataset has been split into a test and a training set using a Leave One Day Out (LOO) approach, where one full day event log is used for testing and remaining days are exploited for training the stochastic model. All experiments were performed with a time bound $t_{max} = 3\,600\,\text{s}$ and a step size $\Delta = 1\,\text{s}$.

4.1 Evaluation on Synthetic Datasets

In order to evaluate the predictor accuracy in a controlled manner, three synthetic datasets are generated through the simulation of a stochastic model of activities. As depicted in Fig. 4, the model is shaped as a semi-Markov process with 4 activities, i.e., $\mathcal{A} = \{\text{act}_0, \text{act}_1, \text{act}_2, \text{act}_3\}$, where act_3 represents the target critical activity.

On completion of any activity, a random switch $p_{x,y}$ makes a selection to establish if the activity sequence must move back to the initial state act_0 or continue with the next activity. The random switch $p_{x,y}$ is set equal to 0.5 for all choices, which represents a kind of worst case for predictability. In so doing, the set of idle states is derived as follow: $\mathcal{I} = \{\text{idle}_{0,0}, \text{idle}_{0,1}, \text{idle}_{1,0}, \text{idle}_{1,2},$

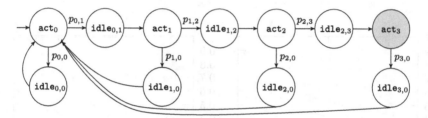

Fig. 4. The stochastic model of activities used for generating synthetic datasets.

$\texttt{idle}_{2,0}, \texttt{idle}_{2,3}, \texttt{idle}_{3,0}\}$. In each state $S = \mathcal{A} \cup \mathcal{I}$, events occur with exponentially distributed inter-times with PDF $f_\lambda(t) = \lambda e^{-\lambda t}$, $\lambda = 1$.

For this model, three different configurations are considered, preserving a mean value $\mu = 2$, but with different values of the coefficient of variation of the sojourn time distribution, so as to evaluate the impact of the dispersion measure on the predictor accuracy. In the first configuration, named *8-phase Erlang*, sojourn times are generated as Erlang distributed random variables with shape $k = 8$ and rate $\lambda = 4$, resulting in a coefficient of variation of $1/\sqrt{8}$. The second configuration, named *2-phase Erlang*, preserves the same Erlang distribution type but varying the shape to $k = 2$, the rate to $\lambda = 1$, and the coefficient of variation to $CV = 1/\sqrt{2}$. In the last configuration, named *2-phase Hyper-exp*, a 2-phase hyper-exponential distribution with $CV \approx 1.92$ and parameters $\lambda_1 = 1/6$, $\lambda_2 = 3/2$, $p_1 = 1/4$, and $p_2 = 3/4$ is applied. For each configuration, a synthetic dataset characterized by a sequence of events and a sequence of activities is generated simulating $4\,000$ completions of sojourn times. In the simulation, the time unit of temporal parameters is 180 s.

Finally, different settings of the predictor component, obtained with the different fitting strategies *Exp*, *Erlang*, and *Whitt* described in Sect. 3 are experimented, so as to evaluate how different fitting distributions impact on prediction and scheduling performance.

Note that, in so doing, the trained model used for prediction is different than that used in the generation of datasets: the *Whitt strategy* will fit expected value and coefficient of variation, while *Exp* and *Erlang strategies* will fit only the expected value; incidentally, for the *2-phase Erlang* dataset, the *Erlang strategy* fits both the coefficient of variation and the overall shape, and so this benevolent case was not reported in the experimentation.

Figure 5 compares precision and recall metrics on the *8-phase Erlang* dataset, for different classes of approximants, for t_d equal to 0, 150, 300, $1\,200$, and $1\,800$ s, and t_w equal to 0, 60, 120, 300 s.

As depicted in Fig. 5b, d and f, increasing the response duration t_d improves the recall for all fitting strategies. Whereas, the qualitative trend of the precision metric varies with the fitting strategy: in the *Whitt strategy*, the highest precision score is reached around 600 s and then steadily maintained until $1\,800$ s, for all t_w except for $t_w = 300\,\text{s}$ (see Fig. 5a); conversely, the *Erlang strategy* showed in Fig. 5c outperforms the precision of the *Whitt strategy* for $t_d \leq 300\,\text{s}$, and

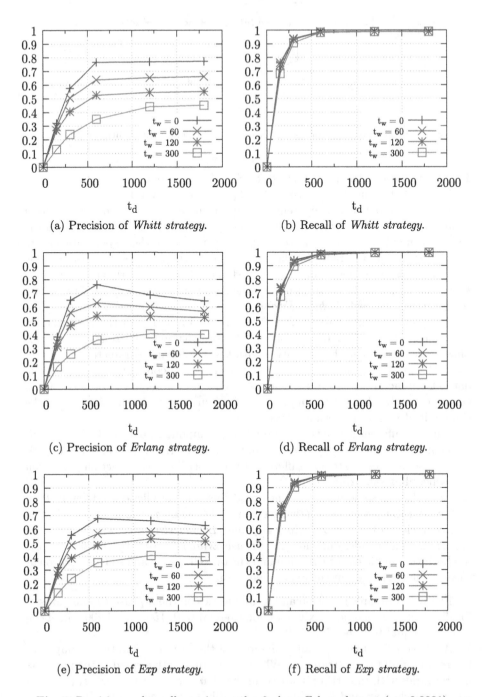

(a) Precision of *Whitt strategy*.

(b) Recall of *Whitt strategy*.

(c) Precision of *Erlang strategy*.

(d) Recall of *Erlang strategy*.

(e) Precision of *Exp strategy*.

(f) Recall of *Exp strategy*.

Fig. 5. Precision and recall metrics on the *8-phase Erlang* dataset ($\epsilon = 0.0001$).

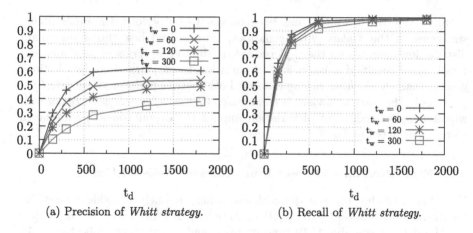

(a) Precision of *Whitt strategy*. (b) Recall of *Whitt strategy*.

Fig. 6. Precision and recall metrics on the *2-phase Erlang* dataset ($\epsilon = 0.0001$).

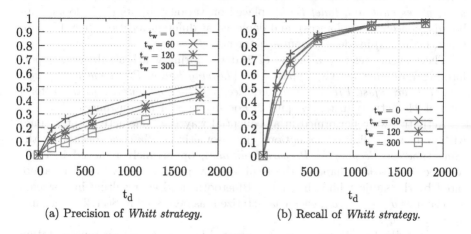

(a) Precision of *Whitt strategy*. (b) Recall of *Whitt strategy*.

Fig. 7. Precision and recall metrics on the *2-phase Hyper-exp* dataset ($\epsilon = 0.0001$).

decays for $t_d \geq 600$ s, resulting in a highest precision score lower than the *Whitt strategy*; finally, the *EXP strategy* (Fig. 5e) follows the same qualitative pattern of the *Erlang strategy*, but with an overall performance score reduced with respect to other cases.

Note that, by increasing the response actuation time t_w (i.e., requiring more time units before the activation of the response action), the overall performance is inevitably reduced, mainly in terms of precision. All metrics are obtained using an ϵ threshold equals to 0.0001. Doubling the ϵ value does not affect precision and recall, while increasing it by a factor of five produces a performance breakdown for $t_d > 600$ s, as the minimum probability required by the scheduling policy is never reached and, consequently, no activation window that contains the start time of the target activity is scheduled.

Since the *Whitt strategy* has been proven to outperform the other fitting strategies on the *8-phase Erlang* dataset, we analyzed how this approximate distribution operates on the *2-phase Erlang* dataset in terms of precision and recall. As a result of a coefficient of variation for sojourn time distributions greater than the *8-phase Erlang* dataset, both metrics perform worse (Fig. 6).

Finally, the same behaviour is emphasized on the *2-phase Hyper-exp* dataset, where a higher coefficient of variation implies the overall lowest performance score, as depicted in Fig. 7.

4.2   Evaluation on a Dataset from Ambient Assisted Living

We experiment the proposed approach also using a publicly available dataset [16] for Activity Recognition (AR) [5,10] of Activities of Daily Living (ADLs) [8].

The dataset contains 1 319 time-stamped and typed events collected by 14 state-change sensors and classified in 14 distinct event types (e.g., *hall-bedroom door*, *plates cupboard*, *toilet flush*), placed on various objects (e.g., doors, cupboards, household appliances), and deployed at different locations (e.g. kitchen, toilet, bedroom) inside a 3-room apartment.

These events refer to a period of 28 days, during which a subject was performing 7 distinct activity types $\mathcal{A} = \{$*Leaving house, Preparing a beverage, Preparing breakfast, Preparing dinner, Sleeping, Taking shower, Toileting*$\}$, for a total of 245 activity instances, plus 219 occurrences of *Idle* instances.

Activities were performed in a sequential way (i.e., one activity at a time) with only some limited exceptions (i.e., sometimes *Toileting* occurs at the same time as *Sleeping* or *Preparing dinner*), opportunely removed in order to be cast into the shape of the semi-Markov process. Activities were annotated by the subject himself using a Bluetooth headset, resulting in a stream of activities $a_1, a_2, \ldots, a_n$, each characterized, as described in Sect. 2, by a tuple $a_i = \langle \alpha_i, \tau_i, \delta_i \rangle$.

As described in [4], events were converted from a *raw* sensor representation (which holds a high signal when the sensor is firing, and a low signal otherwise) into a *dual change-point* sensor representation, which emits a time-point signal when the sensor starts to fire and when it switches off, distinguishing activation/deactivation actions and, consequently, doubling the number of event types and instances, from 14 to 28 and from 1 319 to 2 638, respectively. To avoid some inconsistencies in the characterization of *Leave house*, during which all event types were improperly recorded, we have removed from each training set all events not consistent with the activity.

We consider the case in which the system is required to schedule a response action (e.g., an alarm or a reminder) about the intake of some drug before the beginning of a meal. In this scenario, we perform two distinct experiments where the critical activity is either *Preparing breakfast* or *Preparing dinner*. In both cases, we adopt the *Whitt* strategy for the fitting of sojourn times, which outperformed the other two strategies on synthetic datasets. Results are reported in the first two rows of Table 1, showing precision and recall metrics for different values of t_d and t_w. The prediction of the start of the *Preparing dinner* activity achieves *precision* $= 1, 0.68, 0.68, 0.48$ and *recall* $= 0.806, 0.708, 0.708, 0.6$ for $t_w = 0, 60, 120, 300$ s and $t_d \geq 1200$ s, respectively. For the *Preparing breakfast* activity, prediction performance is considerably worse, resulting in *precision* $= 0.929, 0.462, 0.214, 0.2$ and *recall* $= 0.448, 0.25, 0.136, 0.136$ for $t_w = 0, 60, 120, 300$ s and $t_d \geq 1200$ s, respectively.

An investigation on the structural reasons of reduced performance for *Preparing breakfast* activity provided insight about the existing mismatch between the real process of AAL and the semi-Markov process used in our online prediction. In the real phenomenon captured by the dataset, some activities such as *Toileting* and *Preparing a beverage* occur as a kind of "shuffle" and noisy events between any other two activities, and the subject carries some memory of the previous activity. Whereas, due to the 1-order memory of the SMP abstraction, the process loses any memory of the past history at each start of a new activity.

In principle, this limit could be overcome by adopting a 2-order memory SMP, as already implemented for modelling and distinguishing the *Idle* states on the basis of the previous activity and of the subsequent one. To confirm this conjecture, an additional experiment was carried out by removing all activities of type *Toileting* or *Preparing a beverage* from the dataset. The third row of Table 1 reports *precision* $= 0.672, 0.476, 0.338, 0.329$ and *recall* $= 0.913, 0.896, 0.839, 0.836$ for $t_w = 0, 60, 120, 300$ s and $t_d = 1800$ s, respectively. Precision is thus marginally improved (except for $t_w = 0$ s), while recall greatly improves for all values of t_w.

Table 1. Precision/recall metrics obtained on the real dataset by the *Whitt strategy* for *Preparing dinner* ($\epsilon = 0.0001$), filtered and not-filtered *Preparing breakfast* ($\epsilon = 0.0003$) as critical activities.

			$t_d(s)$					
			0	150	300	600	1200	1800
Preparing dinner	$t_w(s)$	0	0/0	0.163/0.533	0.265/0.684	0.52/0.684	1/0.806	1/0.806
		60	0/0	0.115/0.25	0.2/0.385	0.2/0.385	0.68/0.708	0.68/0.708
		120	0/0	0.2/0.385	0.2/0.385	0.2/0.385	0.68/0.708	0.68/0.708
		300	0/0	0/0	0/0	0/0	0.48/0.6	0.48/0.6
Preparing breakfast	$t_w(s)$	0	0/0	0.093/0.385	0.714/0.385	0.714/0.385	0.929/0.448	0.929/0.448
		60	0/0	0.214/0.136	0.214/0.136	0.214/0.136	0.462/0.25	0.462/0.25
		120	0/0	0/0	0/0	0/0	0.214/0.136	0.214/0.136
		300	0/0	0/0	0/0	0/0	0.2/0.136	0.2/0.136
Preparing breakfast (filtered)	$t_w(s)$	0	0/0	0.465/0.741	0.489/0.759	0.557/0.818	0.661/0.871	0.672/0.913
		60	0/0	0.222/0.556	0.319/0.667	0.33/0.735	0.408/0.828	0.476/0.896
		120	0/0	0.063/0.261	0.128/0.467	0.225/0.69	0.288/0.778	0.338/0.839
		300	0/0	0.053/0.28	0.088/0.429	0.178/0.643	0.277/0.811	0.329/0.836

5 Conclusions

We presented a model-based approach for the online scheduling of response actions, and evaluated its performance both on synthetic and real-world datasets. The results highlighted that sojourn-time statistics provide important information to predict the future evolution, and accurate models based on generally-distributed sojourn times can improve the precision of scheduling policies. When sojourn times are highly variable, or the sequence of activities does not follow regular patterns, scheduling performance suffers from inaccurate predictions. The experimentation on a real-world dataset of AAL confirmed these results and showed that, for AR, future evolution often depends on the history of previous activities, breaking the hypothesis of semi-Markov models. Finally, keeping an idle state between each pair of activities might lead to overfitting the waiting times as the sample size may be small in some case. Performance achieved by different model structures will be investigated in the future.

References

1. Babaoglu, O., Jelasity, M., Montresor, A., Fetzer, C., Leonardi, S., van Moorsel, A.: The self-star vision. In: Babaoğlu, Ö., Jelasity, M., Montresor, A., Fetzer, C., Leonardi, S., Moorsel, A., Steen, M. (eds.) SELF-STAR 2004. LNCS, vol. 3460, pp. 1–20. Springer, Heidelberg (2005)
2. Bencomo, N., France, R., Cheng, B.H.C., Aßmann, U.: Models@run.time: Foundations, Applications, and Roadmaps. Springer, Heidelberg (2014)
3. Bucci, G., Carnevali, L., Ridi, L., Vicario, E.: Oris: a tool for modeling, verification and evaluation of real-time systems. Int. J. SW Tools Technol. Transf. **12**(5), 391–403 (2010)
4. Carnevali, L., Nugent, C., Patara, F., Vicario, E.: A continuous-time model-based approach to activity recognition for ambient assisted living. In: Campos, J., Haverkort, B.R. (eds.) QEST 2015. LNCS, vol. 9259, pp. 38–53. Springer, Heidelberg (2015)
5. Chen, L., Hoey, J., Nugent, C.D., Cook, D.J., Yu, Z.: Sensor-based activity recognition. IEEE Trans. Syst. Man, Cybern. Part C: Appl. Rev. **42**(6), 790–808 (2012)
6. Horváth, A., Paolieri, M., Ridi, L., Vicario, E.: Transient analysis of non-Markovian models using stochastic state classes. Perform. Eval. **69**(7–8), 315–335 (2012)
7. Ihler, A., Hutchins, J., Smyth, P.: Learning to detect events with Markov-modulated Poisson processes. ACM Trans. Knowl. Disc. Data **1**(3), 13 (2007)
8. Katz, S., Downs, T.D., Cash, H.R., Grotz, R.C.: Progress in development of the index of ADL. The Gerontologist **10** (**1 Part 1**), 20–30 (1970)
9. Kephart, J., Chess, D.: The vision of autonomic computing. Computer **36**(1), 41–50 (2003)
10. Kim, E., Helal, S., Cook, D.: Human activity recognition and pattern discovery. IEEE Pervasive Comput. **9**(1), 48–53 (2010)
11. Kulkarni, V.: Modeling and Analysis of Stochastic Systems. Chapman & Hall, Boston (1995)
12. Rasch, K.: An unsupervised recommender system for smart homes. J. Ambient Intell. Smart Environ. **6**(1), 21–37 (2014)

13. Rogge-Solti, A., van der Aalst, W.M.P., Weske, M.: Discovering stochastic petri nets with arbitrary delay distributions from event logs. In: International Business Process Management Workshops, BpPM, pp. 15–27 (2013)
14. Salfner, F., Lenk, M., Malek, M.: A survey of online failure prediction methods. ACM Comput. Surv. **42**(3), 10: 1–10: 42 (2010)
15. Salfner, F., Malek, M.: Using hidden semi-Markov models for effective online failure prediction. In: 26th IEEE International Symposium on Reliable Distributed Systems SRDS 2007, pp. 161–174, October 2007
16. van Kasteren, T., Noulas, A., Englebienne, G., and Kröse, B.: Accurate activity recognition in a home setting. In: Proceedings of International Conference on Ubiquitous Computing, UbiComp 2008, pp. 1–9. ACM, New York (2008)
17. Whitt, W.: Approximating a point process by a renewal process, I: two basic methods. Oper. Res. **30**(1), 125–147 (1982)

Finding Steady States of Communicating Markov Processes Combining Aggregation/Disaggregation with Tensor Techniques

Francisco Macedo[1,2]([✉])

[1] EPF Lausanne, SB-MATHICSE-ANCHP,
Station 8, CH-1015 Lausanne, Switzerland
francisco.macedo@epfl.ch
[2] IST, Alameda Campus, Av. Rovisco Pais, 1, 1049-001 Lisbon, Portugal

Abstract. Stochastic models for interacting processes feature a dimensionality that grows exponentially with the number of processes. This state space explosion severely impairs the use of standard methods for the numerical analysis of such Markov chains. In this work, we develop algorithms for the approximation of steady states of structured Markov chains that consider tensor train decompositions, combined with well-established techniques for this problem – aggregation/disaggregation techniques. Numerical experiments demonstrate that the newly proposed algorithms are efficient on the determination of the steady state of a representative set of models.

1 Introduction

We consider computing the stationary distribution of a continuous–time Markov chain, i.e., solving

$$Ax = 0 \text{ with } \mathbf{1}^T x = 1, \qquad (1)$$

where A is the transpose of the generator matrix of the Markov chain and $\mathbf{1}$ denotes the vector of all ones. Matrix A is non-symmetric, singular and verifies $\mathbf{1}^T A = 0$.

We focus on Markov processes that feature high-dimensional state spaces arising from modelling subsystems that interact with each other. The considered Markov chain consists of d interacting subsystems (processes), and A consequently has a tensor (Kronecker) structure of the form

$$A = \sum_{t=1}^{T} \bigotimes_{k=1}^{d} E_k^t, \qquad (2)$$

where each term in the summation represents a possible transition between states. Problems of this kind are known for their so called state space explosion [6] – exponential growth of the state space dimension on the number of subsystems – which severely impairs the numerical analysis of such Markov processes.

D. Fiems et al. (Eds.): EPEW 2016, LNCS 9951, pp. 48–62, 2016.
DOI: 10.1007/978-3-319-46433-6_4

Applications of these models can be found, e.g., in queueing theory [7,16]; stochastic automata networks [19,26]; analysis of chemical reaction networks [1,20]; or telecommunications [2,25].

Recent algorithms that explore the Kronecker structure in (2) can be found in [3,4,18], where the first is a combination of the other two that intends to avoid the drawbacks of each. They use tensor techniques to solve (1) by using a compressed representation of matrices and vectors to avoid the enormous storage complexity corresponding to the large problem sizes and also to efficiently perform elementary operations, e.g. matrix-vector products. Vectors are, furthermore, approximated. This way, they deal with the state space explosion. The methods mentioned above use tensor train (TT) format, while canonical decomposition (CP) [17] had initially been used in [5]. However, the approximation of vectors inside the algorithms is fundamental and this format does not allow efficient approximations (truncations). This is not the case for TT format as the truncation procedure – TT-SVD algorithm [23] – is based on singular value decomposition (SVD), which is known to have quasi-optimal upper bounds for the error that is done in an approximation.

Aggregation/disaggregation techniques, which can be seen as a subclass of multigrid methods, have also been used to solve (1). A reduced version of (1) where states have been aggregated in groups is solved and then a solution for the original problem needs to be extrapolated using some disaggregation scheme. Aggregation can be defined, for instance: assuming an underlying Kronecker structure of the generator matrix [6,8,15,27]; or imposing specific relations between the rates of transition between states, entries of the generator matrices, of the original and the aggregated processes, in a process called lumpability [9,14].

The approaches proposed in this paper consist of using techniques from the aggregation/disaggregation class that are convenient to adapt to TT format. Two types of operators for aggregation, which result in two proposed algorithms, are considered. Our contribution consists of targeting the computation of the stationary distribution for finite-dimensional communicating Markov processes, developing and comparing different algorithmic approaches, testing their sensitivity to different models by using a broad benchmark collection.

The remainder of this paper is organized as follows. In Sect. 2 we briefly describe tensor-train format. An overview of the ideas behind aggregation/disaggregation algorithms, seen as a subclass of multigrid algorithms, followed by exploring the two mentioned concretizations of interest, is done in Sect. 3. The proposed algorithms, based on combining ideas from the two previous sections, are presented in Sect. 4. In Sect. 5, we analyse the performance of the proposed algorithms for some popular and representative models. Concluding remarks are given in Sect. 6.

2   Tensor Train (TT) Format

We recall the tensor train (TT) format for a vector of length $n_1 n_2 \cdots n_d$, when regarded as an $n_1 \times n_2 \times \cdots \times n_d$ tensor; and of the operator TT format that represents a matrix of sizes $n_1 n_2 \cdots n_d$.

2.1   Representation of a Vector

For a network of d communicating processes, the steady state vector has length $n_1 n_2 \cdots n_d$, where n_μ denotes the number of states in the μ-th process, for $\mu = 1, \ldots, d$. Its entries can be naturally rearranged into an $n_1 \times \cdots \times n_d$ array, defining a d-th order tensor $\mathcal{X} \in \mathbb{R}^{n_1 \times \cdots \times n_d}$, whose entries are denoted by

$$\mathcal{X}_{i_1, i_2, \ldots, i_d}, \quad 1 \le i_\mu \le n_\mu, \quad \mu = 1, \ldots, d.$$

The opposite operation, typically called vectorization, denoted by $\text{vec}(\mathcal{X})$, stacks the entries of $\mathcal{X}$ back into a long vector. The multi-indices of the form $(i_1, i_2, \ldots, i_d)$ are assumed to be traversed in *reverse lexicographical order*. The different n_μ, $\mu = 1, \ldots, d$, are typically called *mode sizes*.

$\mathcal{X}$ becomes a matrix for $d = 2$ and the definition of rank is unique. This rank can be computed using the singular value decomposition (SVD) [12]. The extension of this concept to $d > 2$ is not unique, and thus different notions of low rank decompositions for tensors have been developed – see [17] for an overview.

Tensor train (TT) decomposition [24] benefits from the locality of interactions. A tensor is represented as

$$\mathcal{X}_{i_1, \ldots, i_d} = G_1(i_1) \cdot G_2(i_2) \cdots G_d(i_d), \quad G_\mu(i_\mu) \in \mathbb{R}^{r_{\mu-1} \times r_\mu}, \tag{3}$$

where each $G_\mu(i_\mu)$ is a matrix of size $r_{\mu-1} \times r_\mu$ for $1 \le i_\mu \le n_\mu$. By definition, $r_0 = r_d = 1$. The third order tensors $\mathbf{G}_\mu \in \mathbb{R}^{r_{\mu-1} \times n_\mu \times r_\mu}$, whose slides are the matrices $G_\mu(i_\mu)$, form the building blocks of the TT format and are called the *TT cores*. One expects any explicitly given vector to be possible to write in the form (3), for sufficiently large values r_μ, using the TT-SVD algorithm [23].

Remark 1. Tensor train decomposition is expected to be more suitable for networks with an underlying topology of the processes associated with a train, in the sense that the existing interactions between subsystems should concern consecutive subsystems after they are suitably ordered; recall (3).

If the entries of the tuple $(r_1, r_2, \ldots, r_{d-1})$, called *TT rank*, remain modest, the complexity of storing and performing operations will be significantly reduced. For structured models (networks), its entries are expected to remain small.

2.2   Representation of a Matrix

To efficiently apply a matrix to a vector in TT format, it also needs to be represented in a suitable way. We represent a matrix $A \in \mathbb{R}^{n_1 n_2 \cdots n_d \times n_1 n_2 \cdots n_d}$ using the *operator TT decomposition*

$$A_{(i_1, \ldots, i_d), (j_1, \ldots, j_d)} = A_1(i_1, j_1) \cdot A_2(i_2, j_2) \cdots A_d(i_d, j_d), \tag{4}$$

where $A_\mu(i_\mu, j_\mu) \in \mathbb{R}^{t_{\mu-1} \times t_\mu}$, $t_0 = t_d = 1$ and $(t_1, t_2, \ldots, t_{d-1})$ is the *TT rank*. The tensor $\tilde{\mathcal{X}}$ that results from a matrix-vector product with A, $\text{vec}(\tilde{\mathcal{X}}) = A \cdot \text{vec}(\mathcal{X})$, has a simple TT decomposition. Its new cores, $\tilde{G}_\mu$, are given by

$$\tilde{G}_\mu(i_\mu) = \sum_{j_\mu=1}^{n_\mu} A_\mu(j_\mu, i_\mu) \otimes G_\mu(j_\mu) \in \mathbb{R}^{r_{\mu-1} t_{\mu-1} \times r_\mu t_\mu}, \tag{5}$$

$i_\mu = 1, \ldots, n_\mu$. The entries of the TT rank thus multiply.

3 Aggregation/Disaggregation Schemes

We describe multigrid, as aggregation/disaggregation algorithms are a subclass of them, and then describe particular choices of the restriction (aggregation) operators that should be effective for Markov chains characterized by interacting subsystems.

3.1 Multigrid

Multigrid methods [13] use a set of recursively coarsened representations of the original setting to achieve accelerated convergence.

We describe the core multigrid V-cycle in Algorithm 1.

Algorithm 1. Multigrid V-cycle

1   $v_\ell = \text{MG}(b_\ell, v_\ell)$
2   **if** *coarsest grid is reached* **then**
3 | solve coarse grid equation $A_\ell v_\ell = b_\ell$.
4   **else**
5 | Perform ν_1 smoothing steps for $A_\ell v_\ell = b_\ell$ with initial guess v_ℓ
6 | Compute the residual $r_\ell = b_\ell - A_\ell v_\ell$
7 | Restrict $b_{\ell+1} = Q_\ell r_l$
8 | Restrict $A_{\ell+1} = Q_\ell A_\ell P_\ell$
9 | $v_{\ell+1} = 0$
10 | $e_{\ell+1} = \text{MG}(b_{\ell+1}, v_{\ell+1})$
11 | Interpolate $e_\ell = P_\ell e_{\ell+1}$
12 | $v_\ell = v_\ell + e_\ell$
13 | Perform ν_2 smoothing steps for $A_\ell v_\ell = b_\ell$ with initial guess v_ℓ
14 **end**

The main ingredients that characterize this class of algorithms are: the smoothing scheme; the set of coarse variables; the transfer operators (restriction and interpolation operators); the coarse grid operator.

A more detailed description can be found in [30].

Summing up the main ideas in Algorithm 1 in words: on the way down, at level ℓ, the method performs a certain number ν_1 of smoothing steps, using an iterative solver; the residual of the current iterate is computed and restricted by a matrix-vector multiplication with the restriction matrix for level ℓ, $Q_\ell \in \mathbb{R}^{m_\ell \times m_{\ell+1}}$, where m_ℓ is the number of states at level ℓ; the operator A_ℓ is also restricted via Petrov-Galerkin to get $A_{\ell+1} = Q_\ell A_\ell P_\ell, Q_\ell A_\ell P_\ell \in \mathbb{R}^{m_{\ell+1} \times m_{\ell+1}}$, where $P_\ell \in \mathbb{R}^{m_{\ell+1} \times m_\ell}$ is the interpolation operator at level ℓ; then we have a recursive call where we solve the coarse grid equation, which is the residual equation, in the last level; then, on the way up the grids, the error is interpolated and again some smoothing iterations are applied.

3.2   Restriction and Interpolation on Aggregation/Disaggregation Methods

What separates aggregation/disaggregation techniques [29] from general multi-grid algorithms is that restriction is done in a particular way, associated with aggregation (while interpolation is then associated with disaggregation). A partition of the set of variables is defined and each set of this partition is then associated with one coarse variable.

From the proposed ways to aggregate the states, we focus on two variants in which aggregation (restriction) has a simple Kronecker representation. This is the case when assuming existence of subsystems inside the network.

Tensorized Algorithm Pairing States in Each Subsystem. Aggregation can be done in a tensorized way as proposed in the numerical experiments of [15]. In each subsystem, the states are merged in pairs. This method should be thus particularly suitable for networks for which the states are ordered in each subsystem, associated with a 1D topology.

The simple Kronecker representation of aggregation is strongly related with the fact that aggregation is tensorized. In fact, the associated matrix is a simple Kronecker product of smaller matrices. Each small matrix is associated with one subsystem. Such matrix is, for an example where the number of states of the subsystem of interest is 4,

$$\begin{bmatrix} 1 & 1 & 0 & 0 \\ 0 & 0 & 1 & 1 \end{bmatrix}^T . \tag{6}$$

In particular, defining m_ℓ as the number of states at level ℓ, $m_{\ell+1} = m_\ell/2^d$ as the number of states per subsystem is divided by 2.

The way states are aggregated is represented in Fig. 1, for an example with $d = 2$ and again assuming 4 states per subsystem.

The suitability of a setting with an underlying ordering of the states in each subsystem, associated with a 1D topology, is clear in Fig. 1.

Aggregating All States from Fixed Subsystems. Aggregation can be also done, again assuming the existence of interacting subsystems, through aggregating all states of one particular subsystem in each level [6,8].

Aggregation is again represented with a simple Kronecker product of smaller matrices. Such matrices are now all identity except the one for the dimension

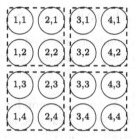

Fig. 1. Example of how aggregation works for $d = 2$ and 4 possible states per subsystem.

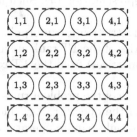

Fig. 2. Example of how aggregation works for $d = 2$ and 4 possible states per subsystem, assuming that the first dimension is the one associated with the subsystem whose states are aggregated.

associated with the subsystem whose states are aggregated. For that particular dimension, we have a column vector of ones. In particular, $m_{\ell+1} = m_\ell/n_i$, where i is the dimension associated with the subsystem whose states are aggregated from level ℓ to $\ell + 1$.

The way states are aggregated is represented in Fig. 2, again for an example with $d = 2$ and 4 states per subsystem.

Two states are aggregated from one level to the next if they are in the same states in a subset of the subsystems that includes all except one, the one whose states are aggregated. In particular, any local topology is now ignored. In Fig. 2, the subsystem whose state is not relevant for the aggregation is the first.

Remark 2. Figure 2 clearly shows how, when the mode sizes start to increase, the distance between states that are aggregated together increases, in case the local topology is associated with ordered states – 1D topology.

4   Proposed Algorithms

In terms of the transpose of the generator matrix of the Markov chain A, the computation of the steady state requires the solution of (1). We focus on problems where the matrix A has the form (2).

4.1   First Variant of the Algorithm from Sect. 3.2 in TT format

Our first proposed algorithm combines the first variant of the algorithm from Sect. 3.2 with TT format by having all involved structures, in particular matrices and vectors, in this format.

Restriction and Interpolation. As the corresponding aggregation is represented as a simple Kronecker product of small matrices, as seen in Sect. 3.2, it allows a TT representation with all entries of the TT rank equal to 1. The relationship between having a simple Kronecker representation and having a simple TT representation is clear. The small matrices from the Kronecker representation are A_μ, $\mu = 1, ..., d$, in the TT representation (4). Such matrices are of the form (6) (adapted to the mode sizes).

We use the transpose of restriction for interpolation, which again has a TT representation where the entries of the TT rank are all 1. In fact, we just need to transpose each of the cores (matrices) from the TT representation of restriction.

Smoother. The tensor structure of the problem rules out smoothers that require access to individual entries of the operator, e.g., Jacobi or Gauss-Seidel. Good candidates are smoothers that only require operations that can be cheaply done within the format. Krylov subspace methods are thus good candidates. We use GMRES, following [4].

Coarse Grid Solver. Coarse grid is still affected by curse of dimensionality as the mode sizes get reduced from one level to another but not the number of dimensions. This is the same type of tensorized multigrid scheme that is used in the scheme proposed in [4], which in particular also suffers from this problem. For that case, a solution, which we adopt in this algorithm, was proposed in [3]. The idea is to use AMEn [10,11] as coarse grid solver.

Implementation Details

Smoother. For smoothing, we use three steps of GMRES in the finest grid while one step in the remaining grids.

Normalization. In (1) we have the restriction that the sum of the entries of the solution is 1. This is not naturally kept during a cycle. We thus normalize the obtained approximation after each cycle.

Truncations. Truncation is needed during a cycle to prevent excessive rank growth. We use the already mentioned TT-SVD algorithm. Looking at Algorithm 1, the TT rank increases in all steps, including inside the smoother, except when applying interpolation and restriction. Truncation is performed after each of these steps.

r_ℓ is truncated with constant accuracy 10^{-1}; the truncation of v_ℓ uses an adaptive scheme where the target accuracy depends on the residual after the previous cycle, being this value times a constant, 10; the accuracy of the truncation of v_ℓ is also used for the truncations inside the GMRES smoother.

A restriction on the maximum rank that one can obtain after each truncation is needed to avoid excessive rank growth. This restriction, originally proposed in [4], depends on the cycle, starting with value 15 and increasing by a factor of $\sqrt{2}$ after cycles for which the variation of the residual norm is smaller than a factor of 0.9 to avoid stagnation.

Size of the Coarse Grid Problem. By construction of restriction and interpolation, the mode sizes in the coarse grid can only be powers of 2. Tests show that mode sizes 4 are already too large so that we consider mode sizes 2.

4.2   Second Variant of the Algorithm from Sect. 3.2 in TT format

The second proposed algorithm combines the second variant of the algorithm discussed in Sect. 3.2 with TT format by again considering the algorithm with all structures in this format.

The comments concerning restriction and interpolation, and also the smoothing procedure, are just the same as for the previous algorithm; see Sect. 4.1.
Coarse Grid Solver. While in the algorithm from Sect. 4.1 the aim of the multigrid scheme is to reduce the mode sizes from one level to another, maintaining the original value of d; in this one we maintain the mode sizes but reduce d. AMEn is thus not so suitable and it is replaced with Moore-Penrose pseudoinverse.

Implementation Details. Most implementation details are as in the algorithm from Sect. 4.1.

Size of the Coarse Grid Problem. As the expensive Moore-Penrose pseudoinverse is the coarse grid solver, we define the number of levels as the smallest value for which the number of states on the coarse grid is smaller than 350.

5   Numerical Experiments

We analyse the performance of the algorithms proposed in Sect. 4. The algorithms of reference for comparison, given their effectiveness when compared against previously proposed algorithms and that they are also in TT format, are the algorithms proposed in [3,18]. They are called MultigridAMEn and AMEn, respectively. As for the algorithm proposed in [4], it might be also considered but as it has the same core as MultigridAMEn, differing only in the coarse grid solver, where the one from MultigridAMEn is proposed to avoid the curse of dimensionality that affects the one from the algorithm of interest; as noted in the implementation details of Sect. 4.1; we do not consider it.

All tests were performed in MATLAB version 2013b, using functions from *TT-Toolbox* [22].

The algorithms from Sects. 4.1 and 4.2 are called 'TensorizedAggregation' and 'DimAggregation', respectively.

Throughout all experiments, we stop an iteration when the residual norm $\|Ax\|$ is two orders of magnitude smaller than the residual norm of the tensor of all ones (scaled so that the sum of its entries is one).

All computation times were obtained on a 12-core Intel Xeon CPU X5675, 3.07 GHz with 192 GB RAM running 64-Bit Linux version 2.6.32.

We use n from here, in the sequence of Sect. 2.1, for the value that is considered for all the mode sizes of a given test case as we will assume equal model sizes in all dimensions.

In the tables that follow: 'Time' stands for the computation time, in seconds; 'Rmax' stands for the maximum entry of the TT rank of the approximated solution; 'Iter' stands for the number of iterations that is needed (measured in cycles for all algorithms except for AMEn, for which it is measured in sweeps, see [18]). We use '-' for algorithms that do not converge.

The experiments are done on some models contained in the collection [21], wide and representative collection of models associated with interacting subsystems. We vary n and d; and use, for the remaining input parameters, the natural generalizations of the default ones.

Each of the following subsections is associated with a different model. For more details, see [21], where the corresponding model is easily found as the names given to the models coincide.

5.1   Model **convergingmetab** in [21]

We consider a model [20] from the field of chemical networks which has a topology of interactions that should not suit TT format particularly well, in the sequence of Remark 1.

We consider $n = 4$ and $d = 10$, respectively; see Table 1.

Table 1. Comparison of the algorithms for model **convergingmetab** $- 4^{10} \approx 1.05 \times 10^6$ states.

	Time	Rmax	Iter
MultigridAMEn	37.4	15	13
AMEn	1.8	16	4
TensorizedAggregation	31.0	29	13
DimAggregation	14.9	15	8

AMEn has an excellent performance. It is, in fact, only expected to have problems if the entries of the TT rank vector are large as this would imply a significant increase on the cost of the subproblem that must be repeatedly solved, as discussed in [18].

As for the proposed algorithms, we see that the restriction (aggregation) operator that is used in TensorizedAggregation is effective as it is the only part where it differs from MultigridAMEn.

DimAggregation behaves particularly well in this context with small mode sizes, which is expected as it is only expected to have problems when the mode sizes are large; recall Remark 2.

Globally, the algorithms, all in TT format, seem to be effective even when the underlying topology is not the theoretically most suitable one, associated with Remark 1.

5.2   Model cyclemetab in [21]

We now consider another model [20] from the field of chemical networks, with a topology that is very far from the ideal one for TT format, see Remark 1, as there is a cycle – last and first subsystems also interact. Additionally, this model is reducible, meaning in particular that the steady state is not unique. This is a class of models that algorithms for finding steady states of Markov chains tend to avoid.

Note that while the methods are intended to be applied to irreducible Markov chains, in the reducible case we have connected components on the set of states so that in order to obtain the desired probabilities from the obtained values, one only needs to then normalize them in a way that the probabilities associated with each connected component sum to 1, as, in the end, each connected component is associated with a different irreducible problem. In particular, the probabilities become unique again.

We go further on n and d, one at each time, considering two cases.
First case – $n = 4$ *and* $d = 18$. The results for the first case are in Table 2.

Table 2. Comparison of the algorithms for model cyclemetab – $4^{18} \approx 6.87 \times 10^{10}$ states.

	Time	Rmax	Iter
MultigridAMEn	525.5	80	22
AMEn	418.7	31	6
TensorizedAggregation	82.2	57	17
DimAggregation	80.9	41	10

The proposed algorithms perform globally well despite the mentioned complicated particularities of this model, considering that the large number of dimensions that is now considered leads to a problem size that is larger than the previous ones. The resulting large entries of the TT rank vector lead to a bad performance of AMEn, while MultigridAMEn is also clearly outperformed by the proposed methods.

The proposed algorithms might perform even better in case there was more control on the rank growth, noting that the entries of the TT rank vector could be much smaller by looking at the maximum entry obtained with AMEn. In fact, the restriction on the maximum entries of the TT rank after the different truncations that is imposed on these algorithms, which is described in the implementation details in Sect. 4.1, is not enough to avoid that the entries of the TT rank become too large. This limitation also affects MultigridAMEn.

The fact that the algorithms are effective even when the underlying topology is not suitable is emphasized as the topology from this model is, as noted before, extremely far from the ideal one.

Second case – $n = 32$ *and* $d = 6$. We address the second case in Table 3.

Table 3. Comparison of the algorithms for model cyclemetab – $32^6 \approx 1.07 \times 10^9$ states.

	Time	Rmax	Iter
MultigridAMEn	432.1	80	19
AMEn	1956.5	31	6
TensorizedAggregation	64.2	57	15
DimAggregation	-	-	-

The performance of DimAggregation is strongly affected by increasing the mode sizes, in the sequence of Remark 2. As for AMEn, it is also more suited for problems with a large number of dimensions than for problems with large mode sizes, as the problem is more structured in the first case; recall Remark 1; thus resulting in this particularly poor performance when the mode sizes are increased, even if the global problem size is smaller than in Table 2.

5.3   Model handoff2class1d in [21]

This model [2, 28] – from the telecommunications field – has a topology of interactions that suits TT format well, in the context of Remark 1. Its main particularity is the existence of two distinguishable types of customers.

As the 1D topology of each subsystem, associated with a natural ordering of the states, is lost, algorithms TensorizedAggregation and MultigridAMEn are not suitable.

Also because of the loss of the local 1D topology, Remark 2 does not apply. The distance between states is not as straight-forward in the new underlying topology but it is clear that it is not as in Fig. 2, and, furthermore, that the distance between states is now smaller. We thus expect DimAggregation to perform properly when n grows.

We go even further on the mode sizes and number of dimensions, considering again two cases.

First case – n = 10 and d = 28. The results for the first case are in Table 4.

Table 4. Comparison of the algorithms for model handoff2class1d – 10^{28} states.

	Time	Rmax	Iter
AMEn	4.7	4	3
DimAggregation	149.7	21	4

AMEn and DimAggregation can easily deal with 28 dimensions, which is expected as they are, as noted before, particularly suited for large d.

In the case of AMEn, as the entries of the TT rank vector are very small, this algorithm is very hard to beat.

The reason why the computation time is much worse for DimAggregation is the problem that was mentioned in the context of Table 2 concerning the far too large entries of the TT rank.

Note that the problem size that is being addressed is clearly the largest that was tested – total of 10^{28} states.

Second case – n = 210 and d = 4. We address the second case in Table 5.

Table 5. Comparison of the algorithms for model handoff2class1d – $210^4 \approx 1.94 \times 10^9$ states.

	Time	Rmax	Iter
AMEn	231.4	8	6
DimAggregation	134.7	29	5

Despite the smaller problem size, when compared with Table 4, AMEn behaves worse because of the larger mode sizes that are now considered; same argument as in the comparison of the performances of AMEn in the tables from Sect. 5.2.

We see that DimAggregation can deal with such a value of n. In fact, while mode sizes 32 are too large for the model in Sect. 5.2, recall Table 3, as expected in the sequence of Remark 2, we now get good convergence for mode sizes 210 as the remark does not apply.

Despite not using any information about the (now more complex) topology of each subsystem, DimAggregation seems to perfectly address models with distinguishable customers. Note that it would be even more competitive if the adopted rank adaptivity scheme would not generate too large entries of its TT rank.

5.4 Choosing the Algorithm to Use

We now propose a way to decide which algorithm to use *depending on the type of model.*

AMEn would be the best option if it was not so strongly affected by cases in which the entries of the TT rank become large which is expected to hold in the most interesting case studies in practice as these should involve large problem sizes. In particular, problems with large mode sizes are common and this is clearly enough, as noted before, for AMEn to stop being efficient.

As for MultigridAMEn, while it is clearly a good option as it performs well for a large and representative collection of models as seen in [3], the restriction and interpolation that are chosen seem to be possible to outperform when replacing them with the aggregation and disaggregation operators proposed in TensorizedAggregation – algorithm from Sect. 4.1 – as repeatedly seen in the tables throughout this section.

In the end, AMEn should be dismissed while MultigridAMEn is outperformed by TensorizedAggregation. TensorizedAggregation and DimAggregation are thus the algorithms of reference.

Models with distinguishable customers have to be addressed using DimAggregation. This algorithm has a particularly good performance, even when the mode sizes increase given that the 1D topology of each subsystem is lost so that Remark 2 is not a problem. As Remark 2 applies to models with indistinguishable customers, DimAggregation should not be used for those. This is, however, a context that perfectly suits TensorizedAggregation. In fact, the proposed algorithms are somehow complementary. For each of the two classes in which we have just partitioned the set of models, there is thus one of the algorithms proposed in this paper that perfectly suits it.

6 Conclusion

We have proposed algorithms for approximating steady states for structured Markov chains combining the concepts of aggregation/disaggregation with TT decompositions.

Numerical experiments with stochastic automata networks demonstrate that, for general problem sizes, they can easily outperform two strong recent algorithms proposed in [3,18]. They can, furthermore, perform remarkably well for very high-dimensional problems. The largest problem size that is addressed is 10^{28}; the largest mode sizes are 210; the maximum number of dimensions is 28.

With the proposed algorithms, all models deriving from Markov chains with a simple Kronecker representation should be possible to address. Robustness is, in fact, one of their main strengths. Besides covering very different fields of interest (chemical networks, queueing networks and telecommunications), the tests done in this paper include: a reducible model, for which the performances of the algorithms, both in comparative and absolute terms, seem to be similar to the performances for a general irreducible model; a model with distinguishable customers, which should be particularly challenging but for which we have an algorithm that perfectly deals with it. These two classes of models are extremely important, both because of their range of applications; but also because of the difficulties, in the past, in addressing them. Additionally, models with topologies that are not suitable for applying TT format provide surprisingly good results.

In the context of the mentioned robustness, when choosing the algorithm to use, depending on the type of model, the choice is quite natural, as seen in Sect. 5.4.

One important limitation that these algorithms have is on the control of the rank growth. Even with a restriction on the maximum entries of the TT rank allowed after the different truncations, the entries of the TT ranks can grow far too much, clearly influencing their performance. This should be further investigated.

Acknowledgments. I thank Daniel Kressner (EPF Lausanne) for helpful discussions.

References

1. Anderson, D.F., Craciun, G., Kurtz, T.G.: Product-form stationary distributions for deficiency zero chemical reaction networks. Bull. Math. Biol. **72**(8), 1947–1970 (2010)
2. Antunes, N., Fricker, C., Robert, P., Tibi, D.: Analysis of loss networks with routing. Ann. Appl. Probab. **16**(4), 2007–2026 (2006)
3. Bolten, M., Kahl, K., Kressner, D., Macedo, F., Sokolović, S.: Multigrid methods combined with amen for tensor structured Markov chains with low rank approximation (2016). arXiv preprint arXiv:1605.06246
4. Bolten, M., Kahl, K., Sokolović, S.: Multigrid methods for tensor structured Markov chains with low rank approximation (2014). arXiv preprint arXiv:1412.0937
5. Buchholz, P.: Product form approximations for communicating Markov processes. Perform. Eval. **67**(9), 797–815 (2010)
6. Buchholz, P., Dayar, T.: On the convergence of a class of multilevel methods for large sparse Markov chains. SIAM J. Matrix Anal. Appl. **29**(3), 1025–1049 (2007)
7. Chan, R.: Iterative methods for overflow queueing networks I. Numer. Math. **51**, 143–180 (1987)
8. Dayar, T.: Analyzing Markov Chains using Kronecker Products: Theory and Applications. Springer Science & Business Media, Heidelberg (2012)
9. Derisavi, S., Hermanns, H., Sanders, W.H.: Optimal state-space lumping in Markov chains. Inf. Process. Lett. **87**(6), 309–315 (2003)
10. Dolgov, S.V., Savostyanov, D.V.: Alternating minimal energy methods for linear systems in higher dimensions. Part I: SPD systems. ArXiv e-prints, January 2013
11. Dolgov, S.V., Savostyanov, D.V.: Alternating minimal energy methods for linear systems in higher dimensions. Part II: Faster algorithm and application to non-symmetric systems. ArXiv e-prints, April 2013
12. Golub, G.H., Van Loan, C.F.: Matrix Computations. Johns Hopkins University Press, Baltimore (1996)
13. Hackbusch, W.: Multi-Grid Methods and Applications. Springer, Heidelberg (2003)

14. Hillston, J., Marin, A., Rossi, S., Piazza, C.: Contextual lumpability. In: Proceedings of the 7th International Conference on Performance Evaluation Methodologies and Tools, pp. 194–203. ICST (Institute for Computer Sciences, Social-Informatics and Telecommunications Engineering) (2013)
15. Horton, G., Leutenegger, S.T.: A multi-level solution algorithm for steady-state Markov chains. In: Gaither, B.D. (ed.), Proceedings of the ACM SIGMETRICS Conference on Measurement and Modeling of Computer Systems, pp. 191–200 (1994)
16. Kaufman, L.: Matrix methods for queuing problems. SIAM J. Sci. Statist. Comput. **4**, 525–552 (1983)
17. Kolda, T.G., Bader, B.W.: Tensor decompositions and applications. SIAM Rev. **51**(3), 455–500 (2009)
18. Kressner, D., Macedo, F.: Low-rank tensor methods for communicating Markov processes. In: Norman, G., Sanders, W. (eds.) QEST 2014. LNCS, vol. 8657, pp. 25–40. Springer, Heidelberg (2014)
19. Langville, A.N., Stewart, W.J.: The Kronecker product and stochastic automata networks. J. Comput. Appl. Math. **167**(2), 429–447 (2004)
20. Levine, E., Hwa, T.: Stochastic fluctuations in metabolic pathways. Proc. Natl. Acad. Sci. U.S.A. **104**(22), 9224–9229 (2007)
21. Macedo, F.: Benchmark problems on stochastic automata networks in tensor train format. Technical report, MATHICSE, EPF Lausanne, Switzerland (2015)
22. Oseledets, I.V.: MATLAB TT-Toolbox Version 2.2 (2011). http://spring.inm.ras.ru/osel/?pageid=24
23. Oseledets, I.V.: Tensor-Train decomposition. SIAM J. Sci. Comput. **33**(5), 2295–2317 (2011)
24. Oseledets, I.V., Tyrtyshnikov, E.E.: Breaking the curse of dimensionality, or how to use SVD in many dimensions. SIAM J. Sci. Comput. **31**(5), 3744–3759 (2009)
25. Philippe, B., Saad, Y., Stewart, W.J.: Numerical methods in Markov chain modelling. Oper. Res. **40**, 1156–1179 (1996)
26. Plateau, B., Stewart, W.J.: Stochastic automata networks. In: Computational Probability, pp. 113–152. Kluwer Academic Press (1997)
27. Pultarová, I., Marek, I.: Convergence of multi-level iterative aggregation-disaggregation methods. J. Comput. Appl. Math. **236**(3), 354–363 (2011)
28. Sidi, M., Starobinski, D.: New call blocking versus handoff blocking in cellular networks. Wirel. Netw. **3**(1), 15–27 (1997)
29. Stewart, W.J.: Introduction to the Numerical Solution of Markov Chains. Princeton University Press, Princeton (1994)
30. Trottenberg, U., Osterlee, C., Schüller, A.: Multigrid. Academic Press, Orlando (2001)

Stability Analysis of a $MAP/M/s$ Cluster Model by Matrix-Analytic Method

Evsey Morozov[1,2] and Alexander Rumyantsev[1,2]($\boxtimes$)

[1] Institute of Applied Mathematical Research, Karelian Research Centre of RAS,
11 Pushkinskaya Str., Petrozavodsk 185910, Russia
ar0@krc.karelia.ru
[2] Petrozavodsk State University, Petrozavodsk, Russia

Abstract. In this paper, we study the stability conditions of the multiserver system in which each customer requires a random number of servers simultaneously and a random service time, identical at all occupied servers. We call it cluster model since it describes the dynamics of the modern multicore high performance clusters (HPC). Stability criterion of an $M/M/s$ cluster model has been proved by the authors earlier. In this work we, again using the matrix-analytic approach, prove that the stability criterion of a more general $MAP/M/s$ cluster model (with Markov Arrival Process) has the same form as for $M/M/s$ system. We verify by simulation that this criterion (in an appropriate form) allows to delimit stability region of a $MAP/PH/s$ cluster model with phase-type (PH) service time distribution. Finally, we discuss asymptotic results related to accelerated stability verification, as well as to the new method of accelerated regenerative estimation of the performance metrics.

Keywords: Stability condition · High performance cluster · Map arrivals · Simultaneous service multiserver system

1 Introduction

A major pivot from frequency scaling of a Central Processing Unit (CPU) to the massive use of multicore and multi-CPU architectures [20] caused a rebirth of interest in studying of stochastic models of modern multiserver systems. A separate class of multiserver models that allow a single customer to be served simultaneously by a number of servers is practically motivated by computing systems such as HPC (as well as cloud/distributed computing) containing a huge number of servers working in parallel. Such systems are used by many users, and a queue manager assigns an available (requested) computational resource to each customer (task submitted by the user). According to [21], the class of systems with simultaneous service has two major subclasses: (i) systems with independent service (service times of a given customer are independent), and (ii) systems with concurrent service (service times of a customer are identical at all occupied servers). While for the subclass (i), the stability conditions in an explicit form have been obtained in a number of papers, see, e.g., [3,6],

© Springer International Publishing AG 2016
D. Fiems et al. (Eds.): EPEW 2016, LNCS 9951, pp. 63–76, 2016.
DOI: 10.1007/978-3-319-46433-6_5

the subclass (ii) requires a more delicate analysis. The key feature of the systems (ii) is that there is a possibility to have idle servers and a non-empty queue simultaneously, which significantly complicates the stability analysis. In the work [1] the stationary distribution of class-dependent delay has been obtained by the system point approach, however the stability analysis of a general multiserver system has not been addressed there. The stability condition obtained in [9] requires a numerical solution of a matrix equation of large dimension. In the recent works [2,5], a *two-server system* is investigated by the matrix-analytic method, and the stability condition (earlier stated with no proof in the paper [1]) has been strictly proven. (More on the matrix-analytic method see in [7,8,10,12].) The work [5] deals with exponential distributions, whereas the work [2] extends the stability condition from [5] to the MAP input. The stability criterion of the cluster model with exponential input, *arbitrary number of servers* and arbitrary distribution of the required number of servers has been obtained in [17] by means of the matrix-analytic approach, while a computationally effective verification of the stability criterion has been proposed in [16]. The main contribution of this paper is an extension of the stability criterion to the cluster model with MAP input. Stability analysis is an essential and challenging stage of investigation of a stochastic model, however stability conditions may be of an independent interest. In particular, the stability criterion of the simultaneous service multiserver model can be used (i) for the capacity planning of an HPC at the design/upgrade stage to obtain the lower bound of the capacity that keeps system stable [18]; (ii) to obtain an upper bound of the potential energy saving [18].

The paper is organised as follows. In Sect. 2, we describe $MAP/M/s$ cluster model. Then, in Sect. 3.2, we study in detail the stability condition of a $MAP/M/s$ cluster model. We show that the earlier found stability criterion of $M/M/s$ model (described briefly in Sect. 3.1) holds true for the MAP arrivals as well. In Sect. 3.3 we demonstrate by simulation that (an appropriately modified) stability condition obtained earlier allows to delimit the stability region of the $MAP/PH/s$ model, at least for the considered parameters. Then, in Sect. 4, we discuss asymptotic results, related to an accelerated verification of stability and accelerated regenerative estimation of the performance metrics.

2   Description of the Model

In this section we describe the $MAP/M/s$ cluster model. For more details on MAP input see [7,10].

We consider a FCFS s-server simultaneous service queueing system with input flow driven by a MAP (D_0, D_1) with k states. Customer i occupies N_i servers simultaneously for the same exponential service time S_i (with mean $1/\mu$). Denote

$$\lambda = \theta D_1 \mathbf{1} \tag{1}$$

the fundamental rate of the MAP, where vector θ satisfies the following equations

$$\begin{cases} \theta D = 0, \\ \theta \mathbf{1} = 1, \end{cases} \tag{2}$$

the matrix $D := D_0 + D_1$, and $\mathbf{1}$ is the vector of ones of the corresponding dimension.

We call customer i *class-j* one if $N_i = j$. The sequence $\{N_i\}$ is assumed to be i.i.d. with a given distribution

$$p_j := \mathsf{P}(N = j), \quad j = 1, \ldots, s \quad \left(\sum_{j=1}^{s} p_j = 1 \right). \tag{3}$$

(We omit the serial index to denote a generic element of an i.i.d. sequence.)

Let $\nu(t)$ be the number of customers in the system at instant t, $t \geqslant 0$. Following [22], we call the vector $m(t) = (m_1(t), \ldots, m_s(t))$ a *macrostate*, where $m_i(t)$ is the class of the ith oldest customer in the system (if $\nu(t) < s$, then $m_i(t) := 0$ for $i > \nu(t)$). Let $\varphi(t) \in \{1, \ldots, k\}$ be the phase of the MAP.

Denote the space of macrostates $\mathcal{M} = \{1, \ldots, s\}^s$. For a fixed $m \in \mathcal{M}$, define

$$\sigma(m) := \max \left\{ i : \sum_{j=1}^{i} m_j \leqslant s \right\}$$

the number of customers *being served* in the macrostate m. Note that $\sigma(m) \leqslant m$. Finally, let $\Omega = \mathbb{Z}_+ \times \mathcal{M} \times \{1, \ldots, k\}$, where $\mathbb{Z}_+ := \{1, 2, \ldots\}$. We use the lexicographical order to enumerate the phases $(m, \varphi) \in \mathcal{M} \times \{1, \ldots, k\}$, and use this multi-dimensional index to refer to the components of matrices. As we show below, the process

$$\left\{ \Theta(t) := (\nu(t), m(t), \varphi(t)) \in \Omega;\ t \geq 0 \right\}, \tag{4}$$

is a Quasi-Birth-Death (QBD) process living in Ω, where $\nu(t)$ is called the *level* of the process.

Fix some state (n, m, x) of the process and consider one-step transitions $(n, m, x) \to (n', m', x')$. The following events (transitions) are possible for levels $n > s$:

1. A *change of MAP phase* with no arrivals: $n' = n, m' = m$, and $x' \neq x$.
2. An *arrival* to the system: $n' = n + 1, m' = m$.
3. *Departure of the ith oldest customer*: $n' = n - 1, x' = x$,

$$m'_j = m_j,\ j < i; \quad m'_{j-1} = m_j,\ j = i + 1, \ldots, s,$$

and m'_s is chosen from (3).

The infinitesimal generator of the QBD process $\{\Theta(t)\}$ with a finite number of phases $d := s^s k$ has the following block-tridiagonal form [10]:

$$\begin{pmatrix} B_1 & B_0 & 0 & 0 & \cdots \\ B_2 & A_1 & A_0 & 0 & \cdots \\ 0 & A_2 & A_1 & A_0 & \cdots \\ 0 & 0 & A_2 & A_1 & \cdots \\ \vdots & \vdots & \vdots & \vdots & \ddots \end{pmatrix},$$

where $A_i, i = 0, 1, 2$ are the square matrices of order d, which contain the intensity of transitions of $\{\Theta(t)\}$ caused by events 2, 1 and 3 defined above, respectively. The matrices $B_i, i = 0, 1, 2$ describe the initial conditions (for levels $n \leqslant s$) and in fact are not used in the stability analysis (for more details see [10]).

Recall also the basic property of the infinitesimal generator of the QBD process

$$A1 = 0, \tag{5}$$

where the matrix $A := A_0 + A_1 + A_2$.

3 Stability Analysis

Below we use the Kronecker product $\otimes$ and Kronecker sum $\oplus$, which is defined as $A \oplus B := A \otimes I + I \otimes B$ for the two matrices A, B and the identity matrix I of the appropriate size. We also use the following property of the Kronecker product. Let A, B, C, D be the matrices of such a size, that AC and BD are possible. Then

$$(A \otimes B)(C \otimes D) = (AC) \otimes (BD). \tag{6}$$

The basic result for stability analysis is the following *Neuts condition for ergodicity* of a QBD process with finite number of states (see [10], Theorem 7.2.4, also [7,12]).

Theorem 1 (Latouche, Ramaswami). *Consider an irreducible, continuous-time QBD process and assume that the matrix A is irreducible. Then, the QBD is positive recurrent if and only if*

$$\alpha A_2 1 > \alpha A_0 1,$$

where α is the unique solution of the system

$$\begin{aligned} \alpha A &= 0, \\ \alpha 1 &= 1. \end{aligned} \tag{7}$$

It is null recurrent if $\alpha A_2 1 = \alpha A_0 1$ and transient if $\alpha A_2 1 < \alpha A_0 1$.

3.1 M/M/s Cluster Model

In this case the QBD process (4) becomes two-dimensional, $\Theta(t) := \{\nu(t), m(t)\}$, $t \geqslant 0$, with $d = s^s$ phases. It has been shown in [17], that the QBD process $\{\Theta(t)\}$ has the infinitesimal generator of block-tridiagonal form. Moreover, the intensity of transitions caused by an arrival/departure at levels $\nu(t) > s$ is governed by the square matrices of size $s^s \times s^s$. Let Q_0 be the matrix of transitions caused by an arrival, that is, the component $Q_0(m, m')$ is the intensity of transition from a macrostate m to a macrostate m' (at some instant t). Note that $Q_0(m, m') = \lambda$ for $m = m'$, and 0 otherwise (since an arrival does not change the macrostate). Let matrix Q_2 contain the intensities of transitions caused by a departure.

(The knowledge of the row and weighted column sums of the matrix Q_2, as an alternative to the componentwise definition of Q_2, has allowed to perform the stability analysis in [17].) Finally, a diagonal matrix Q_1 contains the rates of holding times of the process. It was proved in [17], that the solution of the system

$$\gamma(Q_0 + Q_1 + Q_2) = \mathbf{0},$$
$$\gamma\mathbf{1} = 1,$$

has the following form:

$$\gamma_m = \frac{1}{C} \frac{\prod_{i=1}^{s} p_{m_i}}{\sigma(m)}, \quad m \in \mathcal{M}. \tag{8}$$

Note that the vector $\gamma = (\gamma_m, m \in \mathcal{M})$ can be interpreted as an approximation for distribution of the macrostates for high levels of $\nu(t)$, see [7]. The following result has been proved in [17] for $M/M/s$ cluster model.

Theorem 2. *The irreducible continuous-time QBD process $\{\Theta(t)\}$ is positive recurrent if and only if*

$$\rho := \frac{\lambda}{\mu} C < 1, \tag{9}$$

where

$$C = \sum_{m \in \mathcal{M}} \frac{\prod_{i=1}^{s} p_{m_i}}{\sigma(m)}. \tag{10}$$

It is null recurrent if $\rho = 1$ and transient if $\rho > 1$.

3.2 $MAP/M/s$ Cluster Model

Consider the process $\{\Theta(t), t \geqslant 0\}$ with $d = s^s k$ phases, defined in (4). Now we define the matrices A_i explicitly. Indeed, the matrix A_0 corresponds to the arrivals into the system at high levels $\nu(t) > s$. Then the arrivals do not change the macrostate (since the macrostate is defined only by the s oldest customers). However, an arrival may change the MAP-phase $\varphi(t)$ according to the intensity matrix D_1. As a result, we obtain

$$A_0 := \mathbb{O}_{s^s} \oplus D_1 = I_{s^s} \otimes D_1, \tag{11}$$

where $\mathbb{O}_i$ (I_i) is the square zero (identity) matrix of size i. The Eq. (11) means that there is no change in the macrostate component $m(t)$ of the QBD process, and the change in the MAP-phase component $\varphi(t)$ is governed by the intensity matrix D_1.

Consider the matrix A_2, which corresponds to a departure. Since the MAP-phase is not changed at the departure epoch, and the change of a macrostate is still described by the matrix Q_2, then we obtain

$$A_2 := Q_2 \oplus \mathbb{O}_k = Q_2 \otimes I_k. \tag{12}$$

Denote by $\mathbf{1}_{ks^s}$ the ks^s-dimensional vector of ones, and note that $\mathbf{1}_{ks^s} = \mathbf{1}_{s^s} \otimes \mathbf{1}_k$. Then it follows from the properties of matrix Q_2, definition (12) and property (6), that

$$A_2 \mathbf{1}_{ks^s} = (Q_2 \otimes I_k)(\mathbf{1}_{s^s} \otimes \mathbf{1}_k) = \mu\sigma \otimes \mathbf{1}_k, \tag{13}$$

where the m-th component of column vector σ is defined as $\sigma(m)$, and the equality $Q_2 \mathbf{1}_{s^s} = \mu\sigma$ is proved in [17] (see equality (11) there). To explain the equality, we recall that, for each fixed macrostate m, there are exactly $\sigma(m)$ customers being served, with exponential service times (with intensity μ).

The matrix A_1 corresponds to the transitions of QBD process, which do not change the level. There is the only possibility for this transition, namely, the MAP-phase $\varphi(t)$ may change with no changes of the macrostate $m(t)$. The corresponding transition of the MAP-phase is governed by matrix D_0. Define the square matrix $J := \mathrm{diag}(\sigma)$ of order s^s. Then it follows from the balance condition (5) and equality (13), that the matrix A_1 has the following form:

$$A_1 := -\mu J \oplus D_0 = -\mu J \otimes I_k + I_{s^s} \otimes D_0. \tag{14}$$

Then it follows from (11) and (14), that

$$A_0 + A_1 = -\mu J \oplus D. \tag{15}$$

Now we are ready to find a solution of the system of Eq. (7). By (12) and (15), the first equation of the system (7) is equivalent to

$$\alpha(Q_2 \otimes I_k) = \mu\alpha(J \otimes I_k + I_{s^s} \otimes D). \tag{16}$$

Recall (2), (8) and define vector

$$\alpha := \gamma \otimes \theta. \tag{17}$$

It is easily seen (e.g., by property (6)), that α satisfies the second equation of the system (7).

Using (2) and (6), we obtain

$$\mu\alpha(I_{s^s} \otimes D) = \mu(\gamma \otimes \theta)(I_{s^s} \otimes D) = \mu\gamma \otimes \mathbb{O}_k = \mathbb{O}_{ks^s}. \tag{18}$$

Then, using (17), (18), we can rewrite (16) as

$$(\gamma \otimes \theta)(Q_2 \otimes I_k) = \mu(\gamma \otimes \theta)(J \otimes I_k + I_{s^s} \otimes D).$$

By the property (6), it now follows that

$$(\gamma Q_2 \otimes \theta I_k) = \mu(\gamma J \otimes \theta I_k + \gamma I_{s^s} \otimes \theta D).$$

Finally, a simplification of both parts of the last equality yields

$$\gamma Q_2 \otimes \theta = \mu\gamma J \otimes \theta. \tag{19}$$

It was proved in [17], that $\gamma Q_2 = \mu\gamma J$. Then (19) is also true, that is, the vector α defined in (17) is the solution of system (7).

It follows from (11) and (17), that

$$\alpha A_0 \mathbf{1} = (\gamma \otimes \theta)(I_{s^s} \otimes D_1)\mathbf{1}_{ks^s}$$
$$= (\gamma I_{s^s}\mathbf{1}_{s^s}) \otimes (\theta D_1\mathbf{1}_k) = \theta D_1\mathbf{1}_k =: \lambda. \tag{20}$$

Note that the Eqs. (13), (20), (8) together with (2) give

$$\alpha A_2 \mathbf{1} = (\gamma \otimes \theta)(\mu\sigma \otimes \mathbf{1}_k) = \mu\gamma\sigma = \frac{\mu}{C},$$

where the last equality follows since

$$\gamma\sigma = \frac{1}{C} \sum_{m \in \mathcal{M}} \frac{\prod_{i=1}^{s} p_{m_i}}{\sigma(m)} \sigma(m) = \frac{1}{C}.$$

Thus, the statement of Theorem 2 holds for $MAP/M/s$ cluster model by Theorem 1. Indeed it is an expected result, since the evolution of the MAP component of the process Θ is independent on the macrostate. By this reason, the stability analysis of the MAP/M-type model is similar to the analysis of M/M-type model. In this regard we mention the so-called *local poissonification* of the MAP, introduced in [13].

3.3 Simulation of $MAP/PH/s$ Model

In this section we verify by simulation that stability criterion obtained above holds for a more general $MAP/PH/s$ model as well. However, we do not have a strict proof because of considerable difficulties. The most important is that a few customers may simultaneously start service at a departure instant, and it does not allow to directly extend the Markov process $\{\Theta(t)\}$ to $MAP/PH/s$ model. Nevertheless we believe that the proof can be obtained by the matrix-analytic approach, and we leave it for a future research. Below we provide a numerical example to confirm our conjecture. (Indeed, it is confirmed by a series of the numerical experiments for various driving sequences.) Let S_i have a PH distribution (τ, T) with l states (and $l + 1$ being the absorption state). It is assumed that $\tau\mathbf{1} = 1$, implying $S_i > 0$ [7]. Then the result (9) has the same form, where

$$\mu := \left(-\tau T^{-1}\mathbf{1}\right)^{-1} \tag{21}$$

is the service rate of the PH distribution. We set $s = 50$ and observe 5000 customers. First, we consider an underloaded system with $\rho = 0.9$. We generate the interarrival times $\{T_i\}$ by means of a two-phase MAP process with matrices

$$D_0 = \begin{pmatrix} -4.44 & 2 \\ 1 & -1.49 \end{pmatrix}, \quad D_1 = \begin{pmatrix} 2.44 & 0 \\ 0 & 0.49 \end{pmatrix},$$

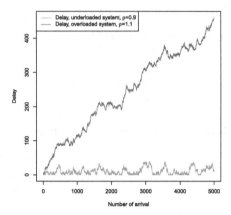

Fig. 1. Comparison of the delays for 5000 customers of a $MAP/PH/s$ cluster model, when the stability condition holds (green), and is violated (red). (Color figure online)

and generate the service times $\{S_i\}$ by a three-phase PH distribution with $\tau = (0.2, 0.4, 0.4)$ and

$$T = \begin{pmatrix} -4 & 2 & 0 \\ 2 & -5 & 1 \\ 1 & 0 & -1 \end{pmatrix}.$$

It then follows, that $\lambda = 1.14$, see (1), and $\mu = 0.909$, see (21). Next, we generate the values $\{p_k\}$ in such a way to obtain $C = 0.717$. It implies $\rho = \lambda C/\mu = 0.9$. To obtain the delay of each customer, we apply the hpcwld package [15] in the R environment [14].

Next, we consider the overloaded system. The interarrival times $\{T_i\}$ follow the two-phase MAP with matrices

$$D_0 = \begin{pmatrix} -4.99 & 2 \\ 1 & -1.6 \end{pmatrix}, \quad D_1 = \begin{pmatrix} 2.99 & 0 \\ 0 & 0.6 \end{pmatrix}.$$

We generate service times $\{S_i\}$ from the same three-phase PH distribution defined above. We also use the same values $\{p_k\}$. Then $\lambda = 1.39$, which implies $\rho = 1.1$.

The resulting delays shown on Fig. 1 demonstrate the (approximate) linear growth of the overloaded system, and a stable behavior of the underloaded system. It indicates that condition (9) allows to delimit the stability/instability zones of the $MAP/PH/s$ cluster model.

4   Some Related Asymptotic Results

4.1   Renewal Approximation for Accelerated Stability Analysis

The calculation of the constant C by Eq. (10) requires significant computational effort, since the power of the set $\mathcal{M}$ equals s^s. By this reason, Eq. (10) is useless in practical cases, since the number of servers s of a contemporary HPC is of the order $10^4 - 10^5$. It has been shown in [16], that the constant C is given by the following expression

$$C = \sum_{i=1}^{s} \frac{1}{i} \sum_{j=i}^{s} p_j^{*i} \sum_{t=s-j+1}^{s} p_t, \qquad (22)$$

where p_j^{*i} is the j-th component of the i-th (discrete) convolution power of vector p, i.e. $p_j^{*i} = P(\hat{N}_1 + \cdots + \hat{N}_i = j)$, where $\hat{N}_i$ are independent copies of a generic variable N.

Note that the summation is done over the number i of customers at service, number j of servers serving customers, number t of servers required by the customer waiting in the head of the queue. Also note that calculation of C by (22) requires $O(s^3 \log s)$ operations.

If we define the renewal process

$$\mathcal{R}(s) = \max \left(k : \sum_{i=1}^{k} \hat{N}_i < s \right),$$

then it follows from (22), that

$$C = \mathsf{E} \frac{1}{\mathcal{R}(s)}.$$

By the Central Limit Theorem (CLT) for a (discrete) renewal process,

$$\mathcal{R}(s) \approx \mathcal{N} \left(\frac{s}{\mathsf{E}N}, \frac{s\mathsf{Var}N}{(\mathsf{E}N)^3} \right),$$

for s large, where $\mathcal{N}$ is the corresponding normal variable. Note that the CLT requires $\mathsf{Var}N < \infty$, that is,

$$\lim_{s \to \infty} \sum_{i=1}^{s} i^2 p_i < \infty.$$

This, in turn, allows to define the following approximation of the constant C:

$$C := C(s) \approx \sum_{i=1}^{s} \frac{1}{i} f_{\mathcal{N}}(i) =: C_0(s), \qquad (23)$$

where the normal density density $f_{\mathcal{N}}(i)$ serves as an approximation of the probability that the r.v. $\mathcal{N}(s/\mathsf{E}N, s\mathsf{Var}N/(\mathsf{E}N)^3)$ belongs to interval $[i-1, i)$. Approximation (23) allows to calculate the constant C in $O(s)$ operations.

To study the accuracy of approximation (23), we perform the following experiment. We assume that distribution $\{p_i\}$ follows the Zipf's law, the discrete analogue of the heavy-tailed Pareto law, widely used in communication and computer systems modeling [4]. Namely,

$$p_i = i^{-z} H_{s,z}^{-1}, \ i = 1, \ldots, s,$$

where constant $z > 1$, and $H_{s,z} := \sum_{i=1}^{s} i^{-z}$ is the generalized harmonic number. Note that if the r.v. N has Zipf's distribution with the exponent z and domain $\{1, \ldots, s\}$, then

$$EN = \frac{H_{s,z-1}}{H_{s,z}},$$

$$VarN = \frac{H_{s,z-2}}{H_{s,z}} - \left(\frac{H_{s,z-1}}{H_{s,z}}\right)^2.$$

Recall also that $\lim_{s \to \infty} H_{s,z} = \zeta(z)$ is the Riemann zeta-function, which is finite only if $z > 1$. Thus the r.v. N has finite variance only if $z > 3$, as $s \to \infty$. For each $z = 1.5, 2.5, 3.5$, we take $s = 10, \ldots, 1000$ and calculate the relative error

$$RE(s) = \frac{C_0(s)}{C(s)} - 1.$$

The results are shown on Fig. 2. As expected, $RE(s) \to 0$ for $z > 3$.

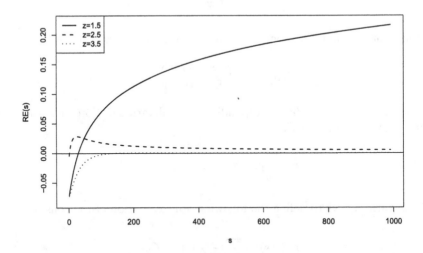

Fig. 2. The relative error $RE(s)$ of the renewal approximation (23) vs. the number of servers $s = 10, \ldots, 1000$, for Zipf's distribution of the number N of servers required by a customer, with exponents $1.5, 2.5, 3.5$.

4.2   Asymptotics of the Limit Probability

The celebrated result for a QBD process $\{X(t), Y(t)\}$ with state space Ω and infinitesimal generator Q under Neuts ergodicity condition is the existence of the limiting probabilities [7]

$$\pi_{i,j} = \lim_{t \to \infty} \mathsf{P}(X(t) = i, Y(t) = j | X(0), Y(0)), \quad (i,j) \in \Omega.$$

The solution $\pi = (\pi_1, \dots)$ of the balance equations $\pi Q = 0, \pi \mathbf{1} = 1$ may be expressed in the matrix-geometric form as

$$\pi_n = \pi_1 R^{n-1}, \tag{24}$$

where $\pi_n = (\pi_{n,1}, \pi_{n,2}, \dots)$; π_1 is the solution of a system of equations corresponding to the boundary states and R is the minimal nonnegative solution of the matrix quadratic equation

$$A_2 R^2 + A_1 R + A_0 = 0.$$

Relation (24) gives, for n large, the following asymtotic result [7]

$$\pi_n \approx (\pi_1 u) v \, (sp(R))^{n-1} + o \, (sp(R))^{n-1}, \tag{25}$$

where u, v are the right and left Perron–Frobenius eigenvectors $(Ru = sp(R)u,$ $vR = sp(R)v)$ corresponding to the Perron–Frobenius eigenvalue $sp(R) < 1$. (The relation $sp(R) < 1$ is guaranteed by the stability condition.) In particular, (25) means that the probability $\mathsf{P}(\nu = k)$ decreases geometrically for large $k > s$.

Recall the well-known classical result for an $M/M/s$ multiserver (classical) system. Denote $\pi_k = \mathsf{P}(\nu = k)$ the probability that the stationary number ν of customers in the system equals k. Then

$$\pi_0 := \left[\sum_{k=0}^{m} \frac{\rho^k}{k!} + \frac{\rho^m}{m!} \frac{m}{m - \rho} \right]^{-1},$$

$$\pi_k := \begin{cases} \pi_0 \rho^k / k!, & 1 \leqslant k \leqslant m, \\ \pi_0 \rho^k / (m! m^{k-m}), & k > m, \end{cases}$$

where $\rho = \lambda/\mu < s$ is the traffic intensity of the multiserver system. It is easy to see that π_k increases for $1 \leqslant k \leqslant \lceil \rho \rceil$ and decreases for $k > \lceil \rho \rceil$. The quantity $\lceil \rho \rceil$ then may be interpreted as the *minimum number of servers that keep the system stable* for a given input rate λ and service rate μ, in this regard see [19]. We also may treat $\lceil \rho \rceil$ as a threshold at which the drift of the workload switches from positive to negative. Indeed, if $\nu < \lceil \rho \rceil$, then the system does not use its full capacity, and the long-run input exceeds the output. On the other hand, if $\nu \geqslant \lceil \rho \rceil$, then the occupied servers are able to process the arriving load, implying negative drift of the workload.

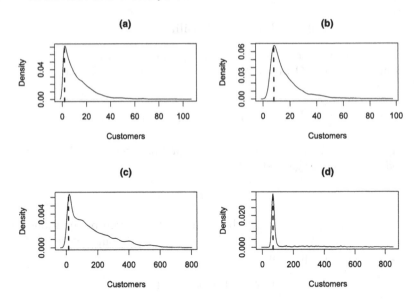

Fig. 3. The mode (dotted vertical line) of the number ν of customers in the system in the $M/M/80$ cluster model, $\lambda = 1, \mu = 1.1C$ and various distributions of the number of servers required by a customer: (a) uniform $p_i = 1/s$, (b) binomial with probability of success $q = 0.1$, $p_i = C_s^i q^i (1 - q)^{s-i}$, (c) bimodal $p_1 = 1 - 1/s, p_s = 1/s$ and (d) bimodal $p_1 = 1 - 1/s^2, p_s = 1/s^2$ distribution.

Surprisingly, the threshold $\lceil \lambda/\mu \rceil$ has a similar interpretation for the cluster model. Indeed, rewrite (9) as

$$\frac{\lambda}{\mu} < \frac{1}{C}.$$

Then $1/C$ is the the mean number of customers that are simultaneously processed in a saturated regime. (Note that $1/C = s$ for a classical multiserver model.) One may also think of $1/C$ as an average number of working servers, where the macrostate is the random environment that switches the servers ON/OFF (at the arrival or departure instants), and each customer requires only one server (instead of N). Then, for $\nu < \lceil \lambda/\mu \rceil$, the system is partially loaded, and the workload has a positive drift. To illustrate this property, we perform the following experiment. We set $s = 80$ and fix arrival intensity $\lambda = 1$. Then we take the following distributions of $\{N_i\}$: (a) uniform, $p_i = 1/s$, (b) binomial, with probability of success $q = 0.1$, implying $p_i = C_s^i q^i (1 - q)^{s-i}$, (c) bimodal, $p_1 = 1 - 1/s, p_s = 1/s$, and (d) bimodal, $p_1 = 1 - 1/s^2, p_s = 1/s^2$. The two latter distributions have large variances. To have $\lambda C/\mu < 1$, we set $\mu = 1.1C$. In all four cases the values C are different. Then we calculate the number of customers in the system at the arrival epochs, for $n = 10^5$ arrivals. We plot the frequency of the number of customers in the system ν, see Fig. 3. In all cases we also plot the maximal value $\lceil \lambda/\mu \rceil$ (dotted line). Figure 3 shows that the value $\lceil \lambda/\mu \rceil$ is the global maximum (mode) of the (empirical) distribution. In particular, the latter result is useful to define the most frequent regenerations for an effective estimation of the performance measures of the cluster model [11].

Acknowledgments. This research is partially supported by Russian Foundation for Basic Research, grants 15-07-02341, 15-07-02354, 15-07-02360, 15-29-07974, 16-07-00622 and the Program of Strategic Development of Petrozavodsk State University. The authors thank Udo Krieger for a few useful comments.

References

1. Brill, P., Green, L.: Queues in which customers receive simultaneous service from a random number of servers: a system point approach. Manag. Sci. **30**(1), 51–68 (1984)
2. Chakravarthy, S., Karatza, H.: Two-server parallel system with pure space sharing and Markovian arrivals. Comput. Oper. Res. **40**(1), 510–519 (2013)
3. Federgruen, A., Green, L.: An M/G/c queue in which the number of servers required is random. J. Appl. Probab. **21**(3), 583–601 (1984)
4. Feitelson, D.G.: Workload Modeling for Computer Systems Performance Evaluation. Cambridge University Press, New York (2015)
5. Filippopoulos, D., Karatza, H.: An M/M/2 parallel system model with pure space sharing among rigid jobs. Math. Comput. Model. **45**(5–6), 491–530 (2007)
6. Gillent, F., Latouche, G.: Semi-explicit solutions for M/PH/1-like queuing systems. Eur. J. Oper. Res. **13**(2), 151–160 (1983)
7. He, Q.-M.: Fundamentals of Matrix-Analytic Methods. Springer, New York (2014)
8. Ibe, O.C.: Markov Processes for Stochastic Modeling. Academic Press, Amsterdam, Boston (2009)
9. Kim, S.S.: M/M/s queueing system where customers demand multiple server use. Ph.D. thesis, Southern Methodist University (1979)
10. Latouche, G., Ramaswami, V.: Introduction to Matrix Analytic Methods in Stochastic Modeling. ASA-SIAM, Philadelphia (1999)
11. Morozov, E., Nekrasova, R., Peshkova, I., Rumyantsev, A.: A regeneration-based estimation of high performance multiserver systems. In: Gaj, P., Kwiecien, A., Stera, P. (eds.) CN 2016. CCIS, vol. 608, pp. 271–282. Springer, Heidelberg (2016). doi:10.1007/978-3-319-39207-3_24
12. Neuts, M.F.: Matrix-Geometric Solutions in Stochastic Models. Johns Hopkins University Press, Baltimore (1981)
13. Neuts, M.F., Dan, L., Surya, N.: Local poissonification of the Markovian arrival process. Commun. Stat. Stoch. Model. **8**, 87–129 (1992)
14. R Foundation for Statistical Computing. Vienna, Austria. ISBN: 3-900051-07-0. http://www.r-project.org/
15. Rumyantsev, A.: hpcwld: High Performance Cluster Models Based on Kiefer-Wolfowitz Recursion. http://cran.r-project.org/web/packages/hpcwld/index.html
16. Rumyantsev, A., Morozov, E.: Accelerated verification of stability of simultaneous service multiserver systems. In: 2015 7th International Congress on Ultra Modern Telecommunications and Control Systems and Workshops (ICUMT), pp. 239–242. IEEE (2015)
17. Rumyantsev, A., Morozov, E.: Stability criterion of a multiserver model with simultaneous service. Ann. Oper. Res., 1–11 (2015). doi:10.1007/s10479-015-1917-2

18. Rumyantsev, A.: An HPC upgrade/downgrade that provides workload stability. In: Malyshkin, V. (ed.) PaCT 2015. LNCS, vol. 9251, pp. 279–284. Springer, Heidelberg (2015)

19. Scheller-Wolf, A., Vesilo, R.: Sink or swim together: necessary and sufficient conditions for finite moments of workload components in FIFO multiserver queues. Queueing Syst. **67**(1), 47–61 (2011)

20. Sutter, H., Larus, J.: Software and the concurrency revolution. Queue **3**(7), 54–62 (2005)

21. Van Dijk, N.M.: Blocking of finite source inputs which require simultaneous servers with general think and holding times. Oper. Res. Lett. **8**(1), 45–52 (1989)

22. Wagner, D., Naumov, V., Krieger, U.R.: Analysis of a multi-server delay-loss system with a general Markovian arrival process. Techn. Hochschule, FB 20, Inst. für Theoretische Informatik, Darmstadt (1994)

Computer Systems and Networking

Predicting Power Consumption in Virtualized Environments

Jóakim von Kistowski[✉], Marco Schreck, and Samuel Kounev

University of Würzburg, Würzburg, Germany
{joakim.kistowski,samuel.kounev}@uni-wuerzburg.de,
marco.schreck@stud-mail.uni-wuerzburg.de

Abstract. Energy efficiency and power consumption of data centers can be improved through conscientious placement of workloads on specific servers. Virtualization is commonly employed nowadays, as it allows for dynamic reallocation of work and abstraction from the concrete server hardware. The ability to predict the power consumption of workloads at different load levels is essential in this context. Prediction approaches can help to make better placement choices at run-time, as well as when purchasing new server equipment, by showing which servers are better suited for the execution of a given target workload. In existing work, power prediction for servers is limited to non-virtualized contexts or it does not take multiple load levels into account. Existing approaches also fail to leverage publicly available data on server efficiency and instead require experiments to be conducted on the target system. This makes these approaches unwieldy when making decision regarding systems that are not yet available to the decision maker. In this paper, we use the readily available data provided by the SPEC SERT to predict the power consumption of workloads for different load levels in virtualized environments. We evaluate our approach comparing predicted results against measurements of power consumption in multiple virtualized environment configurations on a target server that differs significantly from the reference system used for experimentation. We show that power consumption of CPU and storage loads can be reliably predicted with a prediction error of less than 15 % across all tested virtualized environment configurations.

Keywords: Energy efficiency · Power · Prediction · Virtualization · SERT

1 Introduction

Energy efficiency of servers and data centers has become a significant issue over the past decades. In 2010, the U.S. Environmental Protection Agency (U.S. EPA) estimated that 3 % of the entire energy consumption in the U.S. is caused by data center power draw [14]. According to a New York Times study from 2012, data centers worldwide consume about 30 billion watts per hour. This is equivalent to the approximate output of 30 nuclear power plants [1].

© Springer International Publishing AG 2016
D. Fiems et al. (Eds.): EPEW 2016, LNCS 9951, pp. 79–93, 2016.
DOI: 10.1007/978-3-319-46433-6_6

The use of virtualized environments is a common approach to tackle this issue. They abstract the concrete hardware on which work is executed, thus offering more placement options. They can be used to consolidate work on a smaller number of servers or to dynamically change resource allocations at run-time. This may lead to increased energy efficiency despite the initial power and performance overhead caused by virtualization.

Prediction or estimation of workload power consumption is a prerequisite for intelligent placement of work with the goal to optimize energy efficiency. Prediction techniques may be useful at various times. For example, they may help at run-time to predict the effects of workload migration; they may help at deployment time to determine the adequate resource allocation for a new application; finally, they may help when designing a new execution environment to optimize its energy efficiency. The latter is especially challenging as the devices under consideration may not be available yet, preventing instrumentation of these devices.

Prediction of power consumption and energy efficiency of servers must also be capable of dealing with multiple load levels on the devices under consideration. The authors of [8] state that servers usually operate in a load range between 10 % and 50 %. Consequently, while the prediction of power consumption at maximum load may be necessary and useful for capacity planning it offers little information about the actual power draw at run-time.

Existing methods for predicting power consumption that explicitly model virtualized environments are usually designed for run-time prediction using on-line data obtained during system execution. These methods use this run-time data to assist management decisions based on current and past system behavior. The major drawback of such an approach is its inability to provide information about systems that are not yet accessible, meaning that they cannot be used for impact assessment of a virtualized environment during the early stages of a cluster's or cloud's design.

Models that explicitly model the impact of a virtualized environment on the overall energy efficiency and power consumption also do not express differences in target system load levels. They support modeling the power impact of the virtualized environment at maximum load, whereas online models allow for assessment of load levels based on past run-time measurements.

In this paper, we introduce an approach for explicit modeling of the impact a virtualized environment has on power consumption. The approach enable the prediction of this impact on two target load levels of 50 % and 100 %, the former being more relevant for predicting run-time power and the latter being more useful for capacity planning. The method is intended to for off-line power prediction enabling its use before the target device is accessible. To this end, we utilize the data produced by the SPEC Server Energy Efficiency Rating Tool (SERT) [14], used by the U.S. EPA in the Energy Star standard [7].

By using the available SERT results of a target device and performing measurements in a virtualized and non-virtualized environment on an available reference machine, we can predict the impact of virtualization on the power consumption of the inaccessible target device.

The major contributions of this paper are as follows:

1. We measure the power consumption of virtualized environments using the SPEC SERT, showing how standard benchmarks can be used to characterize the energy efficiency of servers running in such environments.
2. We present an approach for predicting the power consumption of a multiple workloads running in a virtualized environment. This approach uses data that can be obtained without having direct access to the target server and enables prediction at two target load levels.

We evaluate our approach by predicting the power consumption of multiple configurations of the Xen hypervisor [5]. We measure virtualized and non-virtualized power consumption using the SPEC SERT, both on a reference (base) machine and on the target system. We predict power consumption of the target system based on measured results of the reference machine and the non-virtualized SERT results of the target system, as they would be measured for the EPA Energy Star submission. We then compare the predicted power consumption of the virtualized environment against our measurements on the target machine.

In the evaluation, we show that our approach is able to predict power consumption reliably for multiple hypervisor configurations at both load levels of 50 % and 100 %. We also show that it features great accuracy for the prediction of CPU-heavy SERT workloads' power consumption.

The remainder of this paper is structured as follows: Sect. 2 describes related work in the area of power prediction models for virtualized environments. Section 3 presents our power measurement methodology with a focus on non-virtualized environments in Sect. 3.1 and a focus on virtualized environments in Sect. 3.2. Section 4 details our prediction approach, followed by its evaluation in Sect. 5. Finally, the paper concludes with Sect. 6.

2   Related Work

Many modeling approaches for prediction of power consumption of server workloads exist. To illustrate the difference to our approach, we classify existing approaches in three non-exclusive categories: General server power models, models for virtualized environments, and cluster/cloud level models.

Server power models model the power consumption of a workload running on a physical machine. Models can be coarse grained or fine grained depending on the required accuracy and the available system information [18]. They employ a variety of mathematical modeling approaches, including interpolation [8,12], regression [15,16], or other machine learning methods [4,17]. Models, such as these may be used to model systems that contain a virtualization layer, even though they do not explicitly model it. Some of these methods are used as inspiration for our explicit modeling of the virtualization layer's impact.

Virtual machine power models, such as [3,6,9], use instrumentation to additionally model the power impact of a virtual machine. These models use on-line data, gathered from the target system for detailed characterization.

Thus, these models focus on immediate run-time decisions based on on-line measurements. In contrast, our approach enables power prediction before the target system is available for running the target load.

Cluster/cloud level models are often provided as part of larger power management decision engines, such as [2,20,21]. These models are created to support online decisions on workload placement within a larger group of servers. They capture the impact of potential decisions in terms of the effects and costs of on-line reconfigurations. To model the power efficiency of an executing workload, they either employ explicit virtual machine power models from the previous category or model the power and performance impact of the virtualized environment implicitly.

Our approach differs from the discussed related work in two key aspects:

1. It is designed to leverage available offline data that can be accessed even if the target system is not available. This allows for power prediction of a workload running within a virtualized environment on an inaccessible system, e.g., before having bought such a system.
2. It considers multiple levels of system load and enables predictions for different load levels.

3   Measuring Power Consumption and Energy Efficiency

We measure power consumption and energy efficiency both in a virtualized and non-virtualized setting using the SPEC SERT [14]. We can assume that SERT results for the non-virtualized configuration are available and may be obtained from the system manufacturer, as they are required by the U.S. EPA Energy Start specification [7]. The following Subsect. 3.1 describes the SERT in its standard (non-virtualized) context, followed by a description of how it may be used in a virtualized environment in Subsect. 3.2.

3.1   Measurement Methodology

We measure power consumption and energy efficiency using SPEC's Server Efficiency Rating Tool (SERT) [14]. SERT was developed by SPEC for the U.S. EPA to use as part of the Energy Star program for enterprise servers [7]. Unlike energy efficiency benchmarks, such as JouleSort [19] and SPECpower_ssj2008 [13], SERT is not intended to provide a single performance or energy efficiency score, but instead it is a rating tool that measures multiple workloads at multiple load levels to derive a thorough characterization of the system under test (SUT). To this end, it uses a set of focused micro-workloads called worklets that exercise the processor, memory and storage I/O subsystems, and may be combined into various configurations running serially or in parallel to provide "system" tests as part of a larger workload.

SERT is designed to be executed on an operating system running directly on the SUT's hardware. The only additional abstraction layer is a Java Virtual Machine (JVM) used to execute Java-based workloads. Previous work has established that the JVM has an insignificant impact on power consumption, but may have an impact on the overall energy-efficiency, depending its performance characteristics [11].

Load Levels: In contrast to many other system power characterization approaches, SERT defines the system's load levels using a system-wide and application centric view instead of defining them based on (CPU-)utilization. The reasoning behind this load level definition is that for any workload and a given SUT there is a maximum transaction rate (throughput) that the workload can achieve. At this maximum transaction rate the system also achieves the maximum possible utilization when running the current workload. SERT defines this point of maximum throughput as full load (100 % load level). Further, load levels are defined using the ratio of their current transaction rate vs. the transaction rate at full load. Consequently, 50 % load is the load level at which the transaction rate is half the maximum transaction rate.

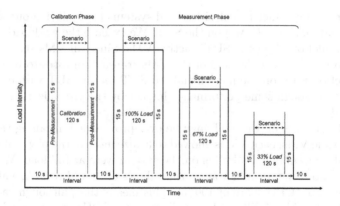

Fig. 1. SERT calibration and measurement phases. [10]

The target load levels are determined using two phases, as depicted in Fig. 1: A calibration phase and a measurement phase. The calibration phase runs the workload at its maximum throughput, determining the 100 % load level. After calibration the measurement phase executes the workload in several intervals. During each interval the inter-arrival waiting time is selected so that the target throughput corresponds to the interval's pre-configured target throughput.

Workloads: SERT features 13 worklets in total: Seven CPU-heavy worklets, two memory-intensive worklets, two storage worklets, one hybrid worklet, and an idle worklet.

The seven CPU worklets are described and analyzed in detail in [11]. Their names are *Compress*, *CryptoAES*, *LU*, *SHA256*, *SOR*, *Sort*, and *XMLValidate*.

All of these worklets are designed with the CPU as their primary bottle-neck resource. Among all of SERT's worklets, the CPU worklets exhibit the greatest power differences between their respective load levels. During a standard measurement each of them is measured at four load levels (25 %, 50 %, 75 %, 100 %). The hybrid worklet (*SSJ*) shares these characteristics with the CPU worklets, it is measured at eight load levels.

The storage and memory worklets are far more stable in their power consumption for different load levels. They are configured to be measured at two load levels (50 % and 100 %) each. The only exception to this is the memory *Capacity* worklet, which features more load levels that are not scaled based on throughput, but rather based on the system's memory capacity. The *Capacity* worklet may also exercise the SUT's storage by emulating swapping behavior. Because of these significant differences to all other worklets, *Capacity* is not used in our prediction approach.

3.2 Measuring Power on Virtualized Systems

We measure energy efficiency of virtualized systems by deploying our workloads within virtual machines (VMs) on the SUT. We execute the workloads' transactions in parallel on all of the SUT's active VMs. Since all VMs share the same physical host, we expect the sum of the VMs transaction rates to be similar to the transaction rates on the non-virtualized SUT (or the SUT with exactly one active VM), although some performance impact by the virtualization itself is to be expected.

Virtual machines on the SUT may be over-provisioned, meaning that in the case of multiple VMs, each VM is allowed to utilize more of the physical resources than would be available if all VMs on the system were at full load. We account for this by also using a configuration that allows each VM to utilize all processors on the SUT (CPU-share of 100 %). This means that, although each VM is theoretically capable of fully utilizing all available CPUs, it may not be able to do so in practice, as other VMs may be occupying some CPU time.

Table 1. Virtual environment configurations.

Config.	# VMs	CPU-share
1	1	100 %
2	2	100 % (over-provisioning)
3	2	50 % (equal sharing)

We measure and predict power consumption and energy efficiency both for virtualized environments that are configured to allow over-provisioning of processing time for the virtual machines, as well as a configuration that shares the available processors equally and does not allow for over-provisioning. Specifically, we use the three configurations shown in Table 1.

Figure 2 shows that our general measurement methodology of achieving load levels by scaling the target throughput works as expected in virtualized environments. Throughput scales linearly over the different load levels with minor differences between the configurations. For the XMLValidate workload, overprovisioning seems to impact performance, resulting in a slightly lower throughput, yet the workload still scales as expected over the load levels.

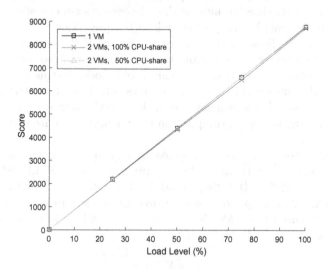

Fig. 2. Throughput-scaling of the different VM configurations for XMLvalidate workload.

4   Prediction Approach

We predict the power consumption of a virtualized environment with a specified configuration on a given target system by using measurements from a reference (base) system. Again, the idea behind our prediction approach is to enable the prediction of power consumption of a virtualized environment on a target machine that is not available to the person making the prediction. Because of this, the only data that we can assume to be available to this person is a standard non-virtualized SERT result of the target machine. The general idea behind our prediction approach is that we train a model to capture how the virtualized environment in its given configuration affects the power characteristics of a system. This training is done using the reference system. Once training is complete, the learned model can be applied to the SERT result of the target system, providing a prediction and characterization of the power impact of the virtualized environment.

To this end, three sets of data must be available. A non-virtualized SERT result of the reference system, a SERT result of the virtualized environment on the reference system, and a SERT result of the non-virtualized target system. The reference system must be available to the person making the prediction, as specific measurements for the target environment must be performed on this system. The SERT results for the systems running without a virtualized environment may be obtained from third party sources (such as the system manufacturer). The non-virtualized results for the reference system can also be obtained through measurements by the person making the prediction.

We use multiple linear regression for prediction of the virtualized environment's impact on power consumption. Using the data obtained for the reference system, we construct a system of equations as our model. In this model, we train the linear coefficients for the regressor variables, which represent the power consumption of the workloads as measured in the non-virtualized environment. For each workload, the power consumption in the virtualized environment serves as response variables.

As stated in Sect. 3, we use 12 workloads in our model, providing us with 12 independent regressor variables: Idle, Compress, CryptoAES, LU, SOR, XML-Validate, Sort, SHA256, HDD Sequential, HDD Random, SSJ, and Flood. Each of these workloads (except Idle) is measured at at least two load levels (50 % and 100 %). We use the workloads at their respective load levels to construct a system of equations as follows (Eq. 1):

$$y = X\beta + \varepsilon \tag{1}$$

where

$$
y = \begin{bmatrix} Idle \\ Compress_{100\%} \\ CryptoAES_{100\%} \\ \vdots \\ Flood_{100\%} \\ Idle \\ Compress_{50\%} \\ CryptoAES_{50\%} \\ \vdots \\ Flood_{50\%} \end{bmatrix}, \beta = \begin{bmatrix} \beta_1 \\ \beta_2 \\ \vdots \\ \beta_{24} \end{bmatrix}, \varepsilon = \begin{bmatrix} \varepsilon_1 \\ \varepsilon_2 \\ \vdots \\ \varepsilon_{24} \end{bmatrix},
$$

and X being the control matrix that contains the regressor variables. X is constructed as in Eq. 2.

$$X = \begin{bmatrix} 0 & Compress_{100\%} & CryptoAES_{100\%} & \dots & Flood_{100\%} \\ Idle & 0 & CryptoAES_{100\%} & \dots & Flood_{100\%} \\ Idle & Compress_{100\%} & 0 & \dots & Flood_{100\%} \\ \vdots & & & \ddots & \vdots \\ Idle & Compress_{100\%} & CryptoAES_{100\%} & \dots & 0 \\ 0 & Compress_{50\%} & CryptoAES_{50\%} & \dots & Flood_{50\%} \\ Idle & 0 & CryptoAES_{50\%} & \dots & Flood_{50\%} \\ Idle & Compress_{50\%} & 0 & \dots & Flood_{50\%} \\ \vdots & & & \ddots & \vdots \\ Idle & Compress_{50\%} & CryptoAES_{50\%} & \dots & 0 \end{bmatrix} \tag{2}$$

However, creating a regression model where each workload is predicted using all other workloads is not optimal. Some workloads' power characteristics differ significantly from other workloads, as they scale differently with load levels or do not scale at all (Idle). This can lead to inaccurate predictions if such workloads are included in an equation for the prediction of another workload for which they are not suited.

We introduce a heuristic for pruning workloads to avoid the use of workloads that may lead to a decrease in prediction accuracy. Specifically, we use the self-prediction error. The self-prediction error is the model's error when predicting the response variables it was trained with. We then create models that only contain subsets of the overall workloads. These sub-models are optimized towards their self-prediction error. This self-prediction error can be minimized either for a single target workload or for all workloads within the sub-model. For a single workload, the self-prediction error is the absolute difference between the predicted and measured power consumption. For all workloads in the sub-model the self-prediction error is the result of Eq. 3.

$$\text{self-prediction error} = \frac{1}{n} \cdot \sum_{i=1}^{n} \left| \frac{\text{predicted Watt}_i - \text{actual Watt}_i}{\text{actual Watt}_i} \right| \tag{3}$$

where i is are the indexes of the n workloads in the sub-model.

For each workload there are $2^{12} = 4096$ possible sub-models, as each workload may or may not be used in the sub-model. Removal of the empty sub-model and sub-models that include only one workload, results in a total of 4083 potential models. Fortunately, as the self-prediction error only depends on the reference system, these combinations must only be tested once and the same sub-models may then be reused for prediction of other target systems.

The sub-models are used for prediction of workload power consumption on the target system. For this prediction, they receive the result of a SERT run on the non-virtualized target system and use this to predict the power consumption the workloads would exhibit on the target system if they ran in the virtualized environment tested on the reference system. An outline of the entire approach is shown in Fig. 3.

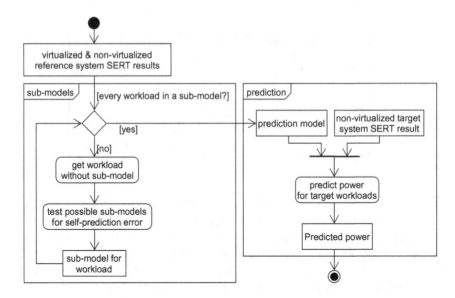

Fig. 3. Power prediction approach outline.

5   Evaluation

We evaluate the accuracy of our prediction approach by predicting the power consumption of a target system and then comparing the predicted consumption against the actually measured power draw. We perform this prediction for multiple workloads and investigate how prediction accuracy is impacted by the target workload choice.

Table 2. Hardware configuration of servers.

	Reference system	Target system
Model	HP ProLiant DL160 Gen9	HP ProLiant DL380 G5
CPU model	Intel Xeon E5-2640 v3	Intel Xeon E5420
Number of cores	8	4
Hardware threads	16	8
CPU frequency	2.60 GHz	2.50 GHz
Memory	2×16 GB	2×8 GB
Harddisk	1×460 GB	1×400 GBA3, 1×120 GB
Network	1 Gbit/s	1 Gbit/s

The systems used in our evaluation have significant differences, as they are from different generations. One system is from 2015 and uses a Haswell generation Intel processor, whereas the other one is an older machine from 2010 with

Table 3. Relative difference between predicted and actual power (no sub-models).

Workload	100 % Load	50 % Load
Idle	487 %	481 %
Compress	97 %	111 %
CryptoAES	95 %	110 %
LU	92 %	108 %
SOR	104 %	116 %
XMLValidate	95 %	109
Sort	98 %	112 %
SHA256	100 %	113 %
Sequential	53 %	55 %
Random	447 %	448 %
SSJ	100 %	113 %
Flood	208 %	203 %

a Harpertown generation processor. More details on the hardware configuration of the systems used for the evaluation are shown in Table 2. These two devices were selected for this evaluation because the power characteristics of the two servers are very different. The reference system has a power consumption in the range between 36.7 W and 141.8 W whereas the target system features an idle power of 237.6 W and a highest measured power consumption of 334.8 W. This means that, because of the significant differences between the devices, there is no intuitive way to guess the power consumption based on simple assumptions and a more complex prediction method is needed.

Both systems run XenServer 6.5 as the hypervisor for their virtualized environments. The virtual machines themselves are running Ubuntu Server (64-bit) 14.04.2 as their guest operating system with Oracle Java Runtime Environment 1.7.0_79. The operating system for the non-virtualized configuration is Debian GNU/Linux 8.0 (Jessie) on the reference system and Ubuntu Server (64-bit) 14.04.2 on the target system.

We measure power consumption of the non-virtualized environment for both systems using the methodology discussed in Sect. 3.1. We then measure the power consumption for each of the three virtualized environment configurations from Table 1 in Sect. 3.2 on both systems. Finally, using the measurement results from the reference system, we predict the power consumption for each of those configurations on the target system.

5.1 Prediction Without Sub-Models

This section investigates if sub-model creation using the self-prediction error as a heuristic is needed. To this end, we predict power consumption using all workloads without any sub-models. Table 3 shows the relative prediction accuracy using this approach. Even using the most accurate regression implementation (stepwise regression for this specific case), the prediction accuracy without sub-model creation is poor. CPU workloads have a relative error around 100 %, with some workloads exhibiting even worse prediction accuracy.

5.2   Self-Prediction Error and Actual Prediction Error

As an intermediate step of the evaluation, we investigate if the self-prediction error of the reference system is actually a good indicator for the final prediction error of a sub-model. The relationship between a sub-model's self-prediction error on the reference system and the actual prediction error regarding the target system for the virtualized environment configuration with one VM is shown in Fig. 4. The figure shows that the self-prediction error is a good indicator for sub-models, as long as the self-prediction error remains low. The great majority of sub-models with self-prediction errors of less than 10 % also has an actual prediction error of less than 10 %.

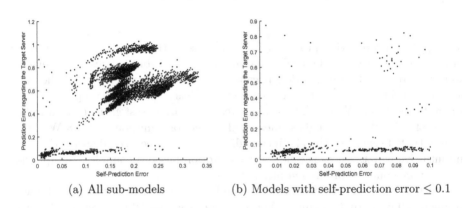

(a) All sub-models   (b) Models with self-prediction error ≤ 0.1

Fig. 4. Relationship between self-prediction error and actual prediction error.

Sub-models with a self-prediction error of more than 10 % are far more volatile and cover a wide range of actual prediction errors. Fortunately, this is not very relevant for our prediction approach, as it always selects the sub-models with the lowest self-prediction error. These observations repeat for the other two virtualized environment configurations.

5.3   Prediction with Sub-models

Finally, we evaluate the prediction error of sub-models obtained using the self-prediction heuristic. We measure the absolute and relative difference between the measured power consumption on the target system and the predicted power. The prediction errors for the virtualized environment configuration with one VM (configuration #1 in Table 1) are shown in Fig. 5.

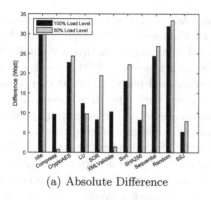
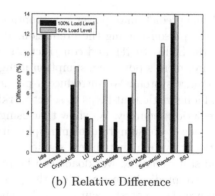

(a) Absolute Difference   (b) Relative Difference

Fig. 5. Prediction errors for configuration with 1 VM.

The CPU-heavy workloads all feature a prediction error of less than 10 %, with Compress and XMLValidate even featuring a relative prediction error of less than 1 % at 50 % load. CryptoAES prediction is the least accurate among the CPU workloads, yet its prediction error does not exceed 10 %. The relative prediction inaccuracy for CryptoAES is easily explained as the workload makes use of the specialized AES CPU instruction set on the reference system. The target system's CPU does not use this instruction set as it is too old.

Idle and the Storage workloads (Sequential and Random) are not as accurate as the CPU workloads with prediction errors between 10 % and 14 %. Storage power prediction is difficult, as the systems have very different storage configurations (1 HDD vs. 2 HDDs) and Idle prediction is hard, since no other workload in this collection resembles it. Under these conditions, the prediction is still quite accurate. The only workload that does not feature any accurate predictions is the Flood workload. It is missing in Fig. 5 because it features a prediction error of 117 % and 118 % for its two load levels. This can be attributed to two factors: The difference in system memory and the absence of any other workload that resembles Flood and that might help to increase its prediction accuracy.

The virtualized environments configurations with two VMs can also be predicted with good accuracy. CPU and storage workloads still remain in the accuracy range of prediction errors less than 15 %. Idle is predicted with slightly diminished accuracy (18 % relative error). The only exception is the CryptoAES workload. Due to its use of a specialized hardware instruction set, it struggles severely with virtual machine over-provisioning (configuration #2 in Table 1). Performance on the reference system itself is already impacted significantly by the virtualized environment configuration. As a result, the prediction accuracy for this configuration suffers and drops to 22 % and 24 %, respectively.

6   Conclusions

This paper introduces an approach for predicting the power consumption of virtualized environments. In contrast to existing approaches, this approach supports off-line prediction of power consumption on a target system that is not available to the person making the prediction. We use the measurement results provided by the SPEC SERT and construct a model that allows predicting a virtualized environment's power consumption using the target system's SERT result and a reference system. We introduce a prediction method based on linear regression and a heuristic that helps to select workloads that have a positive impact on the prediction accuracy. We show that using this heuristic we can accurately predict the power consumption of multiple virtualized environment configurations on a target system that differs significantly from the used reference system. CPU and storage-bound workloads can be predicted with good accuracy, featuring a relative prediction error of less than 15 % on all configurations. CPU workloads feature the greatest accuracy, with some workloads being predicted with errors of less than 1 %. The approach presented in this paper allows for better informed management and cloud/cluster design decisions. Servers might be purchased or requisitioned based on the power consumption the envisioned virtualized environment would cause on these machines.

As part of out future work, we plan to combine the results of our power prediction with the results of existing virtualized environment performance models for energy efficiency prediction. We also plan to extend our model to facilitate prediction for arbitrary third-party workloads.

References

1. Babcock, C.: NY Times data center indictment misses the big picture (2012)
2. Basmadjian, R., Ali, N., Niedermeier, F., de Meer, H., Giuliani, G.: A methodology to predict the power consumption of servers in data centres. In: Proceedings of the 2nd International Conference on Energy-Efficient Computing and Networking-Energy 2011, pp. 1–10. ACM, New York (2011). http://doi.acm.org/10.1145/2318716.2318718
3. Bohra, A.E.H., Chaudhary, V.: VMeter: power modelling for virtualized clouds. In: 2010 IEEE International Symposium on Parallel Distributed Processing, Workshops and Ph.D. Forum (IPDPSW), pp. 1–8, April 2010
4. Chen, J., Li, B., Zhang, Y., Peng, L., Peir, J.K.: Statistical gpu power analysis using tree-based methods. In: 2011 International Green Computing Conference and Workshops (IGCC), pp. 1–6, July 2011
5. Chisnall, D.: The Definitive Guide to the Xen Hypervisor. Pearson Education, Boston (2008)
6. Choi, J., Govindan, S., Urgaonkar, B., Sivasubramaniam, A.: Profiling, prediction, and capping of power consumption in consolidated environments. In: IEEE International Symposium on Modeling, Analysis and Simulation of Computers and Telecommunication Systems, MASCOTS, pp. 1–10, September 2008
7. EPA: ENERGY STAR Program Requirements for Computer Servers, October 2013. https://www.energystar.gov/ia/partners/prod_development/revisions/downloads/computer_servers/Program_Requirements_V2.0.pdf
8. Fan, X., Weber, W.D., Barroso, L.A.: Power provisioning for a warehouse-sized computer. In: The 34th ACM International Symposium on Computer Architecture (2007). http://research.google.com/archive/power_provisioning.pdf

9. Kansal, A., Zhao, F., Liu, J., Kothari, N., Bhattacharya, A.A.: Virtual machine power metering and provisioning. In: Proceedings of the 1st ACM Symposium on Cloud Computing SoCC 2010, pp. 39–50. ACM, New York (2010). http://doi.acm.org/10.1145/1807128.1807136

10. von Kistowski, J., Beckett, J., Lange, K.D., Block, H., Arnold, J.A., Kounev, S.: Energy efficiency of hierarchical server load distribution strategies. In: Proceedings of the IEEE 23nd International Symposium on Modeling, Analysis and Simulation of Computer and Telecommunication Systems (MASCOTS 2015). IEEE, October 2015

11. von Kistowski, J., Block, H., Beckett, J., Lange, K.D., Arnold, J.A., Kounev, S.: Analysis of the influences on server power consumption and energy efficiency for CPU-intensive workloads. In: Proceedings of the 6th ACM/SPEC International Conference on Performance Engineering (ICpPE 2015), NY, USA. ACM, New York, February 2015

12. von Kistowski, J., Kounev, S.: Univariate interpolation-based modeling of power and performance. In: Proceedings of the 9th EAI International Conference on Performance Evaluation Methodologies and Tools (VALUETOOLS 2015), December 2015

13. Lange, K.D.: Identifying shades of green: the SPECpower benchmarks. Computer **42**(3), 95–97 (2009)

14. Lange, K.D., Tricker, M.G.: The design and development of the server efficiency rating tool (SERT). In: Proceedings of the 2nd ACM/SPEC International Conference on Performance Engineering ICPE 2011, pp. 145–150. ACM, New York (2011). http://doi.acm.org/10.1145/1958746.1958769

15. Lee, B.C., Brooks, D.M.: Accurate and efficient regression modeling for microarchitectural performance and power prediction. SIGPLAN Not **41**(11), 185–194 (2006). http://doi.acm.org/10.1145/1168918.1168881

16. Lewis, A., Ghosh, S., Tzeng, N.F.: Run-time energy consumption estimation based on workload in server systems. In: Proceedings of the 2008 Conference on Power Aware Computing and Systems HotPower 2008, p. 4. USENIX Association, Berkeley (2008). http://dl.acm.org/citation.cfm?id=1855610.1855614

17. Niu, D., Wang, Y., Wu, D.D.: Power load forecasting using support vector machine and ant colony optimization. Expert Syst. Appl. **37**(3), 2531–2539 (2010). http://www.sciencedirect.com/science/article/pii/S0957417409007799

18. Rivoire, S., Ranganathan, P., Kozyrakis, C.: A comparison of high-level full-system power models. In: Proceedings of the 2008 Conference on Power Aware Computing and Systems HotPower 2008, p. 3. USENIX Association, Berkeley (2008). http://dl.acm.org/citation.cfm?id=1855610.1855613

19. Rivoire, S., Shah, M.A., Ranganathan, P., Kozyrakis, C.: JouleSort: a balanced energy-efficiency benchmark. In: Proceedings of the 2007 ACM SIGMOD International Conference on Management of Data SIGMOD 2007, pp. 365–376. ACM, New York (2007). http://doi.acm.org/10.1145/1247480.1247522

20. Verma, A., Ahuja, P., Neogi, A.: pMapper: power and migration cost aware application placement in virtualized systems. In: Issarny, V., Schantz, R. (eds.) Middleware 2008. LNCS, vol. 5346, pp. 243–264. Springer, Heidelberg (2008)

21. Verma, A., Ahuja, P., Neogi, A.: Power-aware dynamic placement of HPC applications. In: Proceedings of the 22nd Annual International Conference on Supercomputing ICS 2008, pp. 175–184. ACM, New York (2008). http://doi.acm.org/10.1145/1375527.1375555

Towards Performance Tooling Interoperability: An Open Format for Representing Execution Traces

Dušan Okanović[1], André van Hoorn[1(✉)], Christoph Heger[2], Alexander Wert[2], and Stefan Siegl[2]

[1] Institute of Software Technology, Reliable Software Systems, University of Stuttgart, Stuttgart, Germany
{dusan.okanovic,van.hoorn}@informatik.uni-stuttgart.de
[2] CA Application Performance Management, NovaTec Consulting GmbH, Leinfelden-Echterdingen, Germany
{christoph.heger,alexander.wert,stefan.siegl}@novatec-gmbh.de

Abstract. Execution traces capture information on a software system's runtime behavior, including data on system-internal software control flows, performance, as well as request parameters and values. In research and industrial practice, execution traces serve as an important basis for model-based and measurement-based performance evaluation, e.g., for application performance monitoring (APM), extraction of descriptive and prescriptive models, as well as problem detection and diagnosis. A number of commercial and open-source APM tools that allow the capturing of execution traces within distributed software systems is available. However, each of the tools uses its own (proprietary) format, which means that each approach building on execution trace data is tool-specific.

In this paper, we propose the (OPEN.xTRACE) format to enable data interoperability and exchange between APM tools and (SPE) approaches. Particularly, this enables SPE researchers to develop their approaches in a tool-agnostic and comparable manner. OPEN.xTRACE is a community effort as part of the overall goal to increase interoperability of SPE/APM techniques and tools.

In addition to describing the OPEN.xTRACE format and its tooling support, we evaluate OPEN.xTRACE by comparing its modeling capabilities with the information that is available in leading APM tools.

1 Introduction

Dynamic program analysis aims to get insights from a software system based on runtime data collected during its execution [12]. An important data structure used for dynamic program analysis is the *execution trace*. In its simplest form, an execution trace captures the control flow of method executions for a request served by the system. It can be represented by a dynamic call tree as depicted in Fig. 1a [8]. In the example, the method doGet(..) is the entry point to the processing of a request. The method doGet(..) calls the doFilter(..) method,

© Springer International Publishing AG 2016
D. Fiems et al. (Eds.): EPEW 2016, LNCS 9951, pp. 94–108, 2016.
DOI: 10.1007/978-3-319-46433-6_7

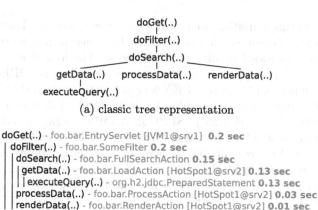

(a) classic tree representation

(b) profiler view

Fig. 1. Example trace

which then calls doSearch(..), etc. The order and nesting of method executions can be obtained by performing a depth-first traversal of the dynamic call tree.

The collection of execution traces is one of the basic features expected from (APM) tools. For instance, it is required to fulfill at least the following three dimensions of APM functionality as defined by Kowall and Cappelli [23]: (i) application topology discovery and visualization, (ii) user-defined transaction profiling, and (iii) application component deep dive. And indeed, the common commercial [1–3,6,7,23] and open-source [18,25] APM tools do support this feature — based on application instrumentation, stack trace sampling, or a mixture of both. It needs to be emphasized that a lot more information than needed for reconstructing dynamic call trees is collected. For instance, the data may include information on timing (e.g., response times, CPU times), variables (HTTP request parameters, SQL queries), or error information (HTTP status code, Java exceptions). Figure 1b shows a simplified profiler-like view on the execution trace from Fig. 1a, including location information and method response times, as provided by most APM tools. However, the type of data and the representation format differ greatly among the different APM tools.

In addition to the aforementioned three APM dimensions, execution traces provide the data set for various further (SPE) activities. For instance, researchers have proposed approaches for extracting and visualizing performance models [11, 16,19], as well as detecting and diagnosing performance problems [17,21,26,27]. Unfortunately, the existing approaches are tailored to the execution trace representations of specific APM tools or custom-made monitoring and tracing implementations.

To summarize, the efficient capturing of detailed execution trace data from software systems during development and operations is widely supported by different APM tools. However, due to diverse and proprietary data formats, approaches building on execution trace data are usually tool-specific.

To overcome this limitation, we propose the Open Execution Trace Exchange (OPEN.XTRACE) format, serving as an open interchange format for representing execution traces provided by different APM tools. The format is accompanied by extensible tooling support to instantiate and serialize the OPEN.XTRACE data, and to import and export OPEN.XTRACE data from/to the data format of leading APM tools. Under the umbrella of the Standard Performance Evaluation Corporation's Research Group (SPEC RG), OPEN.XTRACE is developed as an ongoing community effort among APM/SPE researchers and industry practitioners as a part of the overall goal to increase the interoperability among tools and approaches in this field [28]. The idea of a common format for execution traces goes in line with related community efforts to increase interoperability and usability [30], e.g., for performance [13,24,31] and workload [29] models.

The contribution of this paper is the presentation of the OPEN.XTRACE format, its tooling support, and the evaluation that analyzes the format's completeness by comparing the provided data with the data available in leading commercial and open-source APM tools. It needs to be emphasized that OPEN.XTRACE is a work in progress and that this paper presents the current state.

The remainder of this paper is organized as follows. Section 2 provides an overview of related work. Section 3 describes the OPEN.XTRACE format and its tooling support. Section 4 includes the comparison with APM tools. In Sect. 5, we draw the conclusions and outline future work. Supplementary material for this paper, including the OPEN.XTRACE software and the detailed data supporting the evaluation, is available online [28].

2   Related Work

Related works can be grouped into *(i)* interoperability and exchange formats in software, service, and systems engineering in general, *(ii)* concrete efforts in this direction in performance engineering in particular, as well as into *(iii)* formats for representing trace data.

The idea of having standardized common data formats is not new and not limited to the representation of execution traces. Various efforts in software, service, and systems engineering to provide abstract data models and modeling languages (meta-models) for concrete problems have been proposed and are used in research and practice. Selected examples include TOSCA for representing cloud deployments [9], CIM as an information model for corporate IT landscapes [14], and the NCSA Common Log Format supported by common web and application servers [5]. A well-defined data model (modeling language) comprises an abstract syntax, semantics, and one or more (textual, visual, or a combination of both) concrete syntax [10]. The syntax is commonly based on meta-models, grammars, or schemas (object-relational, XML, etc.). Data models have proven to be most successful if they are developed and maintained by consortia of academic and industrial partners, such as DMTF,[1] OASIS,[2] OMG,[3] or W3C.[4] For this reason, OPEN.XTRACE is being developed as an open community effort driven by SPEC RG from the very beginning.

For workload and performance models, researchers have proposed a couple of intermediate or interchange formats to reduce the number of required transformations between architectural and analytical performance models and tools. KLAPER [13] and CSM (Core Scenario Model) [31] focus on a scenario-based abstraction for performance models, and transformations from/to software design models (e.g., UML SP/MARTE) and analytical models such as (layered) queuing networks are available. Similarly, PMIF (and extended versions of it) [24] focuses on queueing models. WESSBAS [29] is a modeling language for session-based workloads, supporting transformations to different load generators and performance prediction tools.

Few works exist on data formats for execution traces. Knüpfer et al. [22] propose the Open Trace Format (OTF). It is suited for high performance computing, where the most important issues are overhead in both storage space and processing time, and scalability. Although similar in name to our format, OTF is not focused on execution traces, but on collections of arbitrary system events. Similar to OTF, the OpenTracing project provides an API for logging events on different platforms. Unlike OPEN.XTRACE, OpenTracing[5] focuses on so-called spans, i.e., logical units of work—not actual method executions. The Common Base Event format (CBE) was created as a part of IBM's Common Event Infrastructure, a unified set of APIs and infrastructure for standardized event management and data exchange between content manager systems [20]. CBE stores data in XML files. Application Response Measurement (ARM) [15] is an API to measure end-to-end transaction-level performance metrics, such as response times. Transaction-internal control flow is not captured and a data model is not provided. To summarize, there is no open and common format for representing execution traces. The existing formats either represent high-level events or are tailored to specific tools. Section 4 will provide details on execution trace data and representation formats of selected APM tools.

3   OPEN.XTRACE

In Sect. 3.1, we provide an example to introduce additional concepts and terminology. In Sect. 3.2, the main components of OPEN.XTRACE's data model are described in form of a meta-model [10]. In Sect. 3.3, the OPEN.XTRACE instance of the example trace is presented. Section 3.4 presents the tooling support.

---

[1] https://www.dmtf.org/standards.

[2] http://www.oasis-open.org/standards.

[3] http://www.omg.org/spec/.

[4] http://www.w3.org/TR/.

[5] http://opentracing.io/.

3.1   Example and Terminology

The example execution trace shown in Fig. 2, which extends the trace from Fig. 1, results from a HTTP request to a distributed Java enterprise application, whose execution spans over multiple network nodes.

```
1   doGet(..) - foo.bar.EntryServlet ... JVM1@srv1
2     doFilter(..) - foo.bar.SomeFilter
3       doSearch(..) - foo.bar.FullSearchAction
4         getData(..) - foo.bar.LoadAction ... HotSpot1@srv2
5           log(..) - foo.bar.Logger
6           loadData(..) - foo.bar.LoadAction
7             list(..) - foo.bar.ListAction
8               executeQuery(..) - org.h2.jdbc.PreparedStatement
9               executeQuery(..) - org.h2.jdbc.PreparedStatement
10              executeQuery(..) - org.h2.jdbc.PreparedStatement
11        processData(..) - foo.bar.ProcessAction ... JVM1@srv1
12          processSingle(..) - foo.bar.ProcessAction
13          processSingle(..) - foo.bar.ProcessAction
14          processSingle(..) - foo.bar.ProcessAction
15        renderData(..) - foo.bar.RenderAction
```

Fig. 2. Sample trace (textual representation)

The trace starts with the execution of the EntryServlet.doGet() method in the virtual machine *JVM1* on the node *srv1* (line 1). After the initial processing on this node, the execution moves to the node *srv2* (line 4). On this node, after logging an event (line 5), data is fetched from a database by performing several database calls (lines 8–10). Since the database is not instrumented, there are no executions recorded on the node hosting it. After these calls, the execution returns to *srv1* (line 11), where the final processing is performed and the execution ends.

The complete list and structure of method executions to process the client request is denoted as a *trace*. A part of the trace that is executed on a certain *location* is called a *subtrace*. Locations can be identified with server addresses or names, virtual machine names, etc.

Each execution within a trace is called *callable*. This example shows several kinds of executions: method-to-method call (e.g., line 2), move of the execution to a different node (e.g., line 4), logging call (line 5), call to a database (e.g., line 9), and HTTP call (line 1).

A trace contains subtraces and has one subtrace—the one where the execution starts—that acts as a root. Each subtrace can have child subtraces, and it acts as a parent to them. Also, subtraces contain callables, with one callable—the entry point of that subtrace—acting as a root. Callables that can call other callables, e.g., method-to-method calls and remote calls, are called *nesting callables*.

Additionally, each of these entities contains performance-relevant information such as timestamps, response times, and CPU times, which will be detailed in the next section.

3.2 Meta-Model

The core meta-classes of the OPEN.XTRACE model are presented in Fig. 3.

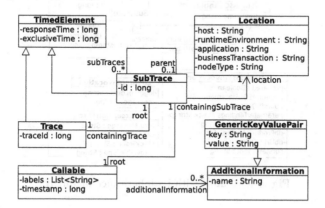

Fig. 3. Trace, SubTrace, Callable and Location

Trace is the container entity that encapsulates an entire execution trace. A **Trace** subsumes a logical invocation sequence through the target system potentially passing multiple system nodes, containers, or applications.

Location specifies an execution context within the trace. It consists of the host identifier, the identifier of the runtime container (e.g., JVM) where the subtrace is executed, the identifier of the application, the business transaction identifier, and the node type. The business transaction specifies the business purpose of the trace. Node type describes the role of the node that the subtrace belongs to, e.g., "Application server" or "Messaging node".

A **SubTrace** represents an extract of the logical **Trace** that is executed within one **Location**.

A **Callable** is a node in a **SubTrace** that represents any callable behavior (e.g., operation execution). For each subtrace there is a root **Callable**, and each **Callable** has its containing subtrace. **AdditionalInformation** can be used to add information on a **Callable** that is tool-specific and not explicitly modeled by OPEN.XTRACE. For simple types of additional information, the **labels** attribute can be used.

Trace and **SubTrace** are extending the **TimedElement** which provides response time and exclusive time. Response time represents the time it takes for an instance to execute. Exclusive time is the execution duration of the instance excluding the execution duration of all nested instances, e.g., a subtrace without its subtraces. If an instance has no nested elements, its exclusive time is equal to its response time.

The detailed inheritance hierarchy of the `Callable` is shown in Fig. 4.

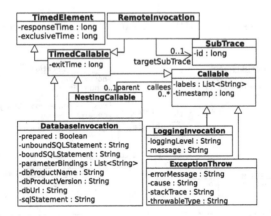

Fig. 4. `Callable` with its inheritance hierarchy

`LoggingInvocation` and `ExceptionThrow` are used for logging and exception events, respectively. `LoggingInvocation` contains information on the logging level and the message, while `ExceptionThrow` contains the message, the cause, and the stack trace of the exception, as well as the class of the exception thrown.

`TimedCallable` is used for modeling exit time for synchronous events that have it, such as method executions and database calls. It also extends the `TimedElement`.

`RemoteInvocation` is used if the execution of the trace moves from one location to another. It points to another `SubTrace`.

Calls to a database are represented with the `DatabaseInvocation`. As we do not expect the monitoring of internals of the database management systems, calls to databases cannot have child callables.

For callables that are able to call other callables, such as methods invoking other methods, `NestingCallable` is used. Each `Callable` can have one parent instance of `NestingCallable` type. Root callables in subtraces do not have parent callables. On the other hand, `NestingCallable` can have multiple children, each of which is of instance `Callable`.

The inheritance hierarchy for `NestingCallable` is shown in Fig. 5.

`MethodInvocation` is used for the representation of method executions. It contains information on the method's signature, e.g., method name, containing class and package, return type, and a list of parameter types, as well as their values. The time a method spent executing on a CPU, with or without the time on the CPU of called methods, is represented using the properties `cpuTime` and `exclusiveCPUTime`, respectively.

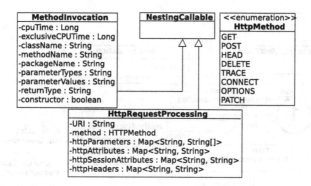

Fig. 5. `NestingCallable` with its inheritance hierarchy

For modeling incoming HTTP calls, `HttpRequestProcessing` is used. It contains the information on URI, HTTP method, parameters, attributes (request and session), and headers. HTTP calls are always the root callables of the subtrace.

In practice, different APM tools provide different sets of data. To avoid situations where we are not sure if some data is missing, or is not supported by a tool, some attributes are marked as optional. For a full list of optional values, please refer to the detailed documentation.

The current version of the OPEN.XTRACE meta-model is implemented in Java [28]. To provide native support for model-driven scenarios, we plan to develop a version using respective technologies such as Ecore [10].

3.3   Model of the Sample Trace

For the trace shown in Fig. 2, the resulting object model would be similar to the model depicted in Fig. 6. The model has been simplified to fit space constraints. Some methods from the listing as well as timing and additional information have been omitted.

The trace model can be read as follows. The execution starts with the *doGet* method (1) on location *srv1*. Other methods are successively called, until the *doSearch* method is called (2). From there, the execution moves to subtrace *subTr2* on location *srv2* (3). After the last method in this subtrace finishes execution (4), the execution returns to *srv1* (2) and continues with the execution of `doSearch` until the end (5).

3.4   Tooling Support

OPEN.XTRACE provides not only the trace meta-model, but also a default implementation, tool adapters, and serialization support, which are publicly available [28].

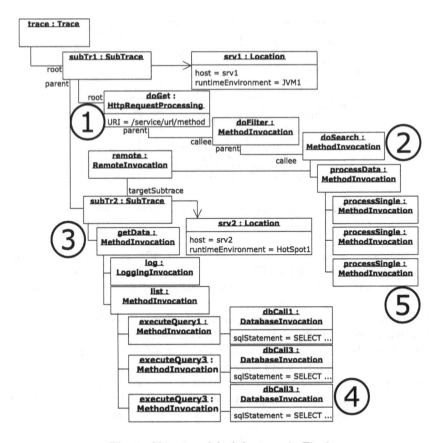

Fig. 6. Object model of the trace in Fig. 2.

Default Implementation. The default implementation of OPEN.xtrace is meant to be used by, e.g., tool developers. Any implementation of the format can be converted into the default implementation and be used as such by the tools.

Adapters. As stated above, in order to translate proprietary trace representations by APM tools into the OPEN.xtrace, adapters are required. Similar to some other well-known approaches (e.g., JDBC), we provide interfaces which are supposed to be implemented by tool vendors or third parties. Currently, we provide publicly available [28] adapters for the following tools: Dynatrace [3], inspectIT [25], Kieker [18], CA APM (previously Wiley Introscope) [2]. Some details on the implementation of each tool adapter include the following:

- Data from **Dynatrace** [3] has to be first exported into the XML format via Dynatrace's session storage mechanism and retrieved via the REST API. After that, the adapter parses the XML file and creates an OPEN.xtrace representation of the traces included in the session storage.

- **inspectIT** [25] stores traces in the form of invocation sequences, in the internal storage called CMR. The adapter connects to the CMR, reads the traces directly from it, and creates the OPEN.XTRACE representation.
- The **Kieker** [18] adapter is implemented as a Kieker plugin. Integrated into the Kieker. Analysis component, it reads traces from a Monitoring Log/Stream, and exports them as traces. Additionally, the adapter supports the reverse direction, i.e., the transformation of OPEN.XTRACE instances into the Kieker format.
- **CA APM** [2] supports exporting of the trace data to XML. As with the Dynatrace, this XML is then parsed, and an OPEN.XTRACE model is created.

Serialization. OPEN.XTRACE provides serialization and deserialization helpers, that are based on the Kryo library.[6] The current implementation provides serialization of binary data, but we also plan to implement a textual serialization. So far, we have not explicitly considered the storage layout and its efficiency.

4 Evaluation

The goal of the evaluation is to assess whether OPEN.XTRACE is expressive enough, i.e., whether it is able to represent the execution trace data provided by common APM tools.

The research questions that we want to answer are as follows:

- *RQ1: Can the APM tools provide the data required for* OPEN.XTRACE?
- *RQ2: Which data available in APM tools are not available in* OPEN.XTRACE?

By investigating RQ1, we want to see what is the level of support for OPEN.XTRACE in available APM tools. The answer to RQ2 will give us the information on the current coverage w.r.t. the modeling coverage and how to further develop OPEN.XTRACE.

We analyzed the data that is provided by APM tools and compared the data they provide with the data that is available in OPEN.XTRACE. Since there are many tools, most of them proprietary, the complete survey of all APM tools is an impossible task. Instead, we focus on the most popular tools, with the largest market share, according to the Gartner report [23]. In our future work, we plan to add information on additional APM tools as a part of the community effort.

The tools and sources of information that we analyzed are as follows.

- **Dynatrace APM** [3]—The trial version, the publicly available documentation, as well as data exported from our industry partners were used.
- **New Relic APM** [6]—The online demo with the sample application was used.However, in the trial version, we were not able to export the data, so the data was gathered from the UI and the available documentation.

---

[6] https://github.com/EsotericSoftware/kryo.

- **AppDynamics APM** [1]—AppDynamics was tested using the trial version.
- **CA APM Solution** [2]—The licensed version of the tool was used to export the traces.
- **Riverbed APM** [7]—The test version with the demo application was used. In this version, we were not able to export the data, so we used the available UI and the documentation.
- **IBM APM** [4]—We used the online demo with the sample application and the provided UI.

Additionally, we analyzed the data from two open source tools: **Kieker** [18] and **inspectIT** [25].

For those tools that did not provide a demo application, the DVD Store[7] was instrumented and used as a sample application. It has to be noted that in this survey we used only the basic distributions of the APM tools. Some of the tools, such as Kieker, have extension mechanisms allowing to measure additional data. For the cases that trial versions were used, to the best of our knowledge, this does not have an influence on the evaluation.

This section presents a condensed overview of the extensive raw data set developed in our study, which is available as a part of the supplementary material [28]. To give an idea of the amount of the investigated APM tool features: the raw table of results includes around 340 features analyzed in each of the eight tools.

Coverage of OPEN.XTRACE. After we collected the data from the tools, we compared the features of OPEN.XTRACE to the data provided by the tools. The comparison is shown in Table 1. The features presented in the rows are extracted from the trace model (Sect. 3.2).

From the table we can see that, while no tool provides all of the data, the method description and timing information is provided by all analyzed tools. The level of detail depends on the tool. IBM is one exception, since their tool provides only aggregated information about method execution over the span of time period. Examples of this kind of data are average, minimal, and maximal response and CPU times, number of exceptions, number of SQL calls, etc.

In other tools, this aggregated data is also available, but this kind of data is of no interest for OPEN.XTRACE, since it is intended to represent single traces.

Data Not Covered by OPEN.XTRACE. The data collected in the survey showed that there is some data that is not covered by OPEN.XTRACE, but is provided by some of the tools. Although this data can be modeled using additional information (see Sect. 3.2), we plan to include it explicitly in our future work.

Synchronization Time, Waiting Time, and Suspended Time. All three mentioned metrics are available in Dynatrace. While OPEN.XTRACE provides means to show that the method was waiting, there are situations

---

[7] http://www.dell.com/downloads/global/power/ps3q05-20050217-Jaffe-OE.pdf.

Table 1. Comparison of data available in OPEN.XTRACE to APM tools

	Metric	OPEN.XTRACE	Kieker	inspectIT	Dynatrace	New Relic	App Dynamics	CA Technologies	Riverbed	IBM
Method description	Method name	•	•	•	•	•	•	•	•	•
	Package name	•	•	•	•	•	•	•	•	
	Class name	•	•	•	•	•	•	•	•	
	Parameter types	•	•	•	•					
	Parameter values	•		•	•					
	Return type	•	•	•						
	Is constructor	•	•	•	•	•	•	•	•	
Timing information	Response time	•	•	•	•	•	•	•	•	•
	Exclusive time	•	•	•	•		•		•	
	Timestamp	•	•	•	•	•	•	•	•	•
	CPU time	•		•	•	•	•			•
	Exclusive CPU time	•			•					
	Exit time	•	•	•	•	•				
Location data	Host	•	•		•	•	•	•	•	•
	Runtime environment	•	•		•	•	•			
	Application	•			•	•	•	•		
	Business transaction	•								
	Node type	•						•		•
Database call information	SQL statement	•		•	•	•	•	•	•	
	Is prepared	•		•	•					
	Bound SQL Statement	•		•	•					
	DB name	•		•		•	•			
	DB version	•		•			•			
	URL	•		•	•	•	•	•	•	
HTTP call information	HTTP method	•		•	•	•		•		
	Parameters	•		•	•		•			
	Attributes	•		•	•					
	Session attributes	•		•	•					
	Headers	•		•	•	•	•		•	
Logging	Logging level	•		•	•					
	Message	•		•	•					
Error information	Error message	•		•	•	•	•	•	•	
	Cause	•		•			•			
	StackTrace	•		•	•		•			
	Throwable type	•		•	•		•			

where it is important to know *why* the method was on hold. Synchronization time represents periods of waiting for access to a synchronization block or a method. Waiting time is the time spent waiting for an external component. Suspended time is the time the whole system was suspended due to some external event during which it could not execute any code.

Nested Exceptions. The nested exception can point to the real cause of the problem and therefore provide valuable information for the analysis. This metric is available in Dynatrace.

Garbage Collector. There is a set of performance issues related to garbage collection, so this information can help to identify them. This metric is available in New Relic, App Dynamics and IBM APM.

Thread Name. There are situations where a certain thread or thread group causes a problem. Adding this information to the location description would make the diagnosis of these problems easier. The thread name metric is available in Dynatrace, New Relic, and CA. The thread group name is available in CA.

HTTP Response Code and Response Headers. Knowing the state of the HTTP response can be important for detecting problems in traces that include HTTP calls. The response code is available in Dynatrace, New Relic, Riverbed, and IBM, while New Relic additionally provides response header contents.

5   Conclusion

Execution trace data is an important basis for different SPE approaches. While a number of commercial and open-source APM tools provides the support for capturing of execution traces within distributed software systems, each of the tools uses its own (proprietary) format.

In this paper we proposed OPEN.xtrace and its tooling support, which provides a basis for execution trace data interoperability and allows for developing tool-agnostic approaches. Additionally, we compared OPEN.xtrace with the information that is available in leading APM tools, and evaluated its modeling capabilities. Our evaluation showed the level of support for the format in most popular APM tools, and provided us with the guidelines on how to further extend the format.

Since this is a community effort, we plan to engage the public, including APM tool vendors to influence the further development of OPEN.xtrace, all under the umbrella of the SPEC RG [28]. Future work includes extensions of the modeling capabilities, e.g., to support asynchronous calls, and to support additional APM tools via respective adapters. In the long term, we want to extend the effort by including also non-trace data, e.g., system-level monitoring data in form of time series data.

Acknowledgements. This work is being supported by the German Federal Ministry of Education and Research (grant no. 01IS15004, diagnoseIT), by the German Research Foundation (DFG) in the Priority Programme "DFG-SPP 1593: Design For Future— Managed Software Evolution" (HO 5721/1-1, DECLARE), and by the Research Group of the Standard Performance Evaluation Corporation (SPEC RG, http://research.spec. org). Special thanks go to Alexander Bran, Alper Hidiroglu, and Manuel Palenga— Bachelor's students at the University of Stuttgart—for their support in the evaluation of the APM tools.

References

1. AppDynamics—Application Performance Monitoring and Management. https:// www.appdynamics.com/
2. CA—Application Performance Management. http://www.ca.com/us/products/ ca-application-performance-management.html
3. Dynatrace—Application Monitoring. http://www.dynatrace.com/en/application-monitoring/
4. IBM—Application Performance Management. http://www.ibm.com/middleware/ us-en/knowledge/it-service-management/application-performance-management. html
5. Logging control in W3C httpd. https://www.w3.org/Daemon/User/Config/ Logging.html
6. New Relic APM. https://newrelic.com/application-monitoring
7. Riverbed—Application Performance Monitoring. http://www.riverbed.com/de/ products/steelcentral/application-performance-management.html
8. Ammons, G., Ball, T., Larus, J.R.: Exploiting hardware performance counters with flow and context sensitive profiling. In: Proceedings of the ACM SIGPLAN 1997 Conference on Programming Language Design and Implementation (PLDI 1997), pp. 85–96 (1997)
9. Binz, T., Breitenbücher, U., Kopp, O., Leymann, F.: TOSCA: portable automated deployment and management of cloud applications. In: Advanced Web Services, pp. 527–549 (2014)
10. Brambilla, M., Cabot, J., Wimmer, M.: Model-Driven Software Engineering in Practice, 1st edn. Morgan & Claypool Publishers, Williston (2012)
11. Brosig, F., Huber, N., Kounev, S.: Automated extraction of architecture-level performance models of distributed component-based systems. In: Proceedings of the 26th IEEE/ACM International Conference on Automated Software Engineering (ASE 2011), pp. 183–192 (2011)
12. Canfora, G., Penta, M.D., Cerulo, L.: Achievements and challenges in software reverse engineering. Commun. ACM **54**(4), 142–151 (2011)
13. Ciancone, A., Drago, M.L., Filieri, A., Grassi, V., Koziolek, H., Mirandola, R.: The KlaperSuite framework for model-driven reliability analysis of component-based systems. Softw. Syst. Model. **13**(4), 1269–1290 (2014)
14. Distributed Management Task Force: Common Information Model (CIM) Standard, February 2014. http://www.dmtf.org/standards/cim/
15. Elarde, J.V., Brewster, G.B.: Performance analysis of application response measurement (ARM) version 2.0 measurement agent software implementations. In: Proceedings of the 2000 IEEE International Performance, Computing, and Communications Conference (IPCCC 2000), pp. 190–198 (2000)

16. Fittkau, F., Finke, S., Hasselbring, W., Waller, J.: Comparing trace visualizations for program comprehension through controlled experiments. In: Proceedings of the 2015 IEEE 23rd International Conference on Program Comprehension (ICPC 2015), pp. 266–276 (2015)

17. Heger, C., van Hoorn, A., Okanović, D., Siegl, S., Wert, A.: Expert-guided automatic diagnosis of performance problems in enterprise applications. In: Proceedings of the 12th European Dependable Computing Conference (EDCC 2016). IEEE (2016, to appear)

18. van Hoorn, A., Waller, J., Hasselbring, W.: Kieker: a framework for application performance monitoring and dynamic software analysis. In: Proceedings of the 3rd ACM/SPEC International Conference on Performance Engineering (ICPE 2012), pp. 247–248 (2012)

19. Israr, T.A., Woodside, C.M., Franks, G.: Interaction tree algorithms to extract effective architecture and layered performance models from traces. J. Syst. Softw. **80**(4), 474–492 (2007)

20. Jacob, B., Lanyon-Hogg, R., Nadgir, D., Yassin, A.: A Practical Guide to the IBM Autonomic Computing Toolkit. IBM, Indianapolis (2004)

21. Kiciman, E., Fox, A.: Detecting application-level failures in component-based internet services. IEEE Trans. Neural Netw. **16**(5), 1027–1041 (2005)

22. Knüpfer, A., Brendel, R., Brunst, H., Mix, H., Nagel, W.E.: Introducing the open trace format (OTF). In: Alexandrov, V.N., Albada, G.D., Sloot, P.M.A., Dongarra, J. (eds.) ICCS 2006. LNCS, vol. 3992, pp. 526–533. Springer, Heidelberg (2006)

23. Kowall, J., Cappelli, W.: Magic quadrant for application performance monitoring (2014)

24. Lladó, C.M., Smith, C.U.: PMIF+: extensions to broaden the scope of supported models. In: Balsamo, M.S., Knottenbelt, W.J., Marin, A. (eds.) EPEW 2013. LNCS, vol. 8168, pp. 134–148. Springer, Heidelberg (2013)

25. NovaTec Consulting GmbH: inspectIT. http://www.inspectit.eu/

26. Parsons, T., Murphy, J.: Detecting performance antipatterns in component based enterprise systems. J. Object Technol. **7**(3), 55–91 (2008)

27. Rohr, M., van Hoorn, A., Giesecke, S., Matevska, J., Hasselbring, W., Alekseev, S.: Trace-context sensitive performance profiling for enterprise software applications. In: Proceedings of the SPEC International Performance Evaluation Workshop (SIPEW 2008), pp. 283–302 (2008)

28. SPEC Research Group: OPEN—APM interoperability initiative (2016). http://research.spec.org/apm-interoperability/

29. Vögele, C., van Hoorn, A., Schulz, E., Hasselbring, W., Krcmar, H.: WESSBAS: extraction of probabilistic workload specifications for load testing and performance prediction–a model-driven approach for session-based application systems. J. Softw. Syst. Model. (2016). Under revision

30. Walter, J., van Hoorn, A., Koziolek, H., Okanovic, D., Kounev, S.: Asking "what"?, automating the "how"?: the vision of declarative performance engineering. In: Proceedings of the 7th ACM/SPEC on International Conference on Performance Engineering, pp. 91–94. ICPE 2016. ACM (2016)

31. Woodside, C.M., Petriu, D.C., Petriu, D.B., Shen, H., Israr, T., Merseguer, J.: Performance by unified model analysis (PUMA). In: Proceedings of the 5th International Workshop on Software and Performance (WOSP 2005), pp. 1–12 (2005)

Feedback in Recursive Congestion Control

David A. Hayes[1]([✉]), Peyman Teymoori[2], and Michael Welzl[2]

[1] Simula Research Laboratory, Fornebu, Norway
davidh@simula.no
[2] University of Oslo, Oslo, Norway
{peymant,michawe}@ifi.uio.no

Abstract. In recursive network architectures such as RINA or RNA, it is natural for multiple layers to carry out congestion control. These layers can be stacked in arbitrary ways and provide more ways to use feedback than before (which of the many controllers along an end-to-end path should be notified?). This in turn raises concerns regarding stability and performance of such a system of interacting congestion control mechanisms. In this paper, we report on a first analysis of feedback methods in recursive networks that we carried out using a fluid model with a packet queue approximation. We find that the strict pushback feedback based on queue size can have stability issues, but robust control can be achieved when each congestion controller receives feedback from all sources of congestion within and below its layer.

1 Introduction

Different forms of feedback have been investigated since the beginning of research on congestion control. It can be implicit (e.g., using packet loss or growing delay as an indication of congestion) or explicit (e.g., using an Explicit Congestion Notification (ECN) bit or multiple bits). From the location where congestion appears, it can propagate forward towards the receiver (from where it is in some way reflected to the sender) or directly backward towards the sender. There have been proposals for hop-by-hop congestion control and *backpressure* mechanisms – some of them are discussed in [6], a survey that was published as early as 1980. We mention ATM as a technology that is notable for supporting many of the explicit feedback methods mentioned above, via Resource Management (RM) cells that allowed explicit rate feedback signalling as well as ECN marking, both in the forward and backward direction.

Recursion has only recently been proposed as an architectural approach to networking that should remove some problems that have been identified with traditional network layering. Examples include the Recursive InterNetwork Architecture (RINA) [3] and the Recursive Network Architecture (RNA) [18]. A key concern about such architectures is their stability when congestion controllers operate concurrently at different layers over various scopes. RNA proposes a

---

D. Hayes — completed most of the work on this paper while with the University of Oslo.

© Springer International Publishing AG 2016
D. Fiems et al. (Eds.): EPEW 2016, LNCS 9951, pp. 109–125, 2016.
DOI: 10.1007/978-3-319-46433-6_8

simple approach to manage this problem for some cases [17]. However, a thorough evaluation is missing in both of RNA and RINA. Despite the multitude of feedback methods covered by the literature on congestion control over the last decades, to the best of our knowledge, how to best provide congestion control feedback in a *recursive* network architecture has never been studied in detail before. We have begun to investigate congestion control for recursive networks with [16]. Using some simple scenarios, this work shows that concatenated/stacked layers ("DIFs") of RINA can improve performance. However, only one type of feedback was considered in these scenarios where every DIF ran a simple TCP-like congestion control with Explicit Congestion Notification (ECN).

In this work we carry out a first investigation of the possibilities for feedback in recursive networks. We do this by means of a fluid-type model that is evaluated using Simulink [14]. We use a sender behavior that is suitable for the experimentation and representative of a modern congestion control mechanism, which makes our findings broadly applicable. Our contribution consists of both new, and perhaps unexpected, findings about congestion control feedback as well as the model itself, which seeks to strike a balance between being simple enough to manipulate and complete enough to allow for meaningful investigations of control loop interactions in recursive networks.

2   Related Work

Before and since the aforementioned survey [6] was published, a huge number of congestion control mechanisms have been devised, in many contexts, using all kinds of feedback along the categories that we have explained. The diversity is such that we point at two textbooks ([8,20]) instead of even trying to provide a reasonably comprehensive set of examples here. In the following, we briefly introduce the two recursive network architectures RINA and RNA instead.

The traditional Internet protocol stack does not match all usage scenarios well. It can become quite complex in practice with Multi-Protocol Label Switching (MPLS), Quality of Service (QoS), proxies and middle boxes. Different layers often perform similar functions of addressing and flow control. RINA was presented in [3] as a way of solving these and many other Internet architecture problems [4,7,9,15]).

In RINA every layer, called a "Distributed InterProcess Communication (IPC) Facility" (DIF), has the same structure and set of "mechanisms". A DIF provides IPC services to its processes over a certain scope. The mechanisms of a DIF can be programmed via "policies", which allows DIFs to be optimized for their operating environment (e.g. wireless links). Among other mechanisms, every DIF has a transport-like protocol called "Error and Flow Control Protocol" (EFCP), and a "Relaying and Multiplexing Task" (RMT) module which represents routing and multiplexing functions. This allows each DIF in the network to detect and manage the congestion of its own resources, pushing back to/signaling higher layer DIFs when resources are overloaded.

RNA [18] adopts a similar recursive approach to RINA with regards layering. It presents a single "reusable" protocol (it is also called "metaprotocol"—analogous to RINA's DIF) for all the layers of the protocol stack. This protocol presents a number of generic services to the other horizontal or vertical layers/protocols. In other words, RNA provides a single protocol that can be tuned at different layers to dynamically support new services and unify all the conventional and overlay layers [17]. This metaprotocol is capable of adopting a variety of communication functions including congestion control.

3   A Model Based Investigation of Feedback for Recursive Congestion Control

Recursive architectures allow congestion control mechanisms to operate on flow aggregates at lower layers as well as end-to-end at the uppermost layer. This potentially allows for more versatile control, however, it also creates complex control interactions between the various loops. We investigate these interactions using a simple, but not trivial, model. Too simple a model will lack the complex control loop interactions that we wish to investigate, too complex a model will not be solvable. We adopt the RINA terminology in our description and discussion of our model, but the principles are applicable to other recursive models such as RNA. Figure 1 shows the RINA topology we model, representing a layered model of how hosts might be connected via Internet Service Providers (ISPs) [13]. All DIFs (RINA terminology for layers) in the model implement congestion control except the lowest peer-to-peer (P2P) DIFs.

In contrast with two well-known non-recursive feedback mechanisms that we use as a baseline—(i) End-to-end forward ECN[1] like feedback, and like feedback, and backward ECN (BECN) (see footnote 1) like feedback — we investigate the comparative stability of three different possible recursive congestion feedback mechanisms: (i) Recursive forward feedback, (ii) Recursive backward feedback, and (iii) Recursive pure pushback feedback. We explain these mechanisms with an example in Sect. 3.1.

3.1   Experimental Setup

In our experiments the router nodes (x-GW and R-n in Fig. 1) are the bottlenecks in the end to end path, each with the capacity to carry 100 rate units. Packet queue models are limited to a maximum of 100 packets in these tests. Test traffic flows from the sender to the receiver with a propagation delay between each node in the network of 10 ms. Delay between DIFs at the same node is modelled as 0 ms.

Apart from the traffic flowing end-to-end, cross traffic traverses each of the router nodes. The amount of cross traffic changes every second, drawn from a uniform random distribution between 0 and 10 % of the capacity (average 5 % of capacity).

---

[1] In our experiments congestion is a measure of queueing above a threshold, so richer than standard Internet ECN.

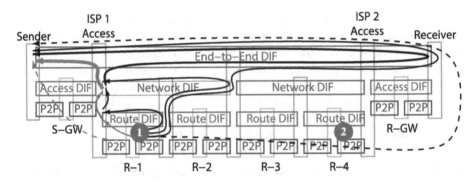

Fig. 1. Feedback mechanisms being investigated: Non-recursive congestion control with (i) Forward ECN like feedback (black dashed arrow), (ii) Congestion control with Backward ECN like feedback (light blue dashed arrow); Recursive congestion control with: (iii) Backward feedback (light blue arrows), (iv) Forward feedback (black arrows). (v) Pure pushback feedback follows a similar path to (iii), but only carries congestion information from the layer below— in this model each DIF can only infer congestion from queue growth related to congestion in the DIF below. (Color figure online)

To investigate the effect sudden disturbances have on the stability of the system, we introduce a cross traffic of 40 % of the bottleneck capacity, first at location ① ($t = [50, 100]$ s) and then later at ② ($t = [150, 200]$ s). ① was chosen to be relatively close to the sender and ② close to the receiver to highlight the ·difference between the backward and forward feedback mechanisms.

Figure 1 shows the feedback message paths for the mechanisms being investigated using disturbance ① as an example. In our experiments we feedback a congestion signal based on queue size. This contains more information than the binary ECN signal used in the Internet and the pure pushback signal currently used in RINA, however, the aim in this initial work is to compare each feedback mechanism at its best.

Non-recursive Feedback. These mechanisms are used as a base to compare the recursive RINA feedback mechanisms with. In these scenarios the end points are connected by routers with no intermediate layers. Only end-to-end congestion control operates between the sender and the receiver. Figure 1 shows the Internet like non-recursive feedback paths as dashed lines with arrows. The black dashed line depicts the path of standard forward ECN signals, while the lighter blue dashed lines with arrows depicts the path of the BECN signals. The model uses queue size as the congestion signal, a richer signal than generally used in the Internet, but better for comparison.

Recursive Feedback. The solid lines depict two possible recursive RINA congestion notification paths. The black lines with arrows depict forward end reflection type notifications. These are similar to forward ECN, except that they operate on each affected DIF in the layered architecture. The lighter blue lined

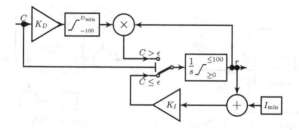

Fig. 2. Sender model. Input: congestion signal C. Output: send rate r.

arrows depict back propagating notifications, similar in some sense to BECN, but operating at each layer. As the congestion signals move up the layers they carry congestion information from all of the layers below them (indicated in the diagram by thicker lines). Both the recursive forward and backward mechanisms use pushback congestion notifications in addition to their explicit notifications.

The pure pushback feedback mechanism allows a very neat demarcation between the layers, however, it does not allow the overarching feedback loops common in system control applications [11,12]. Pure pushback congestion notification relies on the lowest DIF closest to the congestion reacting and reducing its send rate. This in turn causes the DIF above to react to queue growth and reduce its send rate. The process repeats, pushing the congestion notification back all the way to the sender (if necessary). This process may involve signalling within a DIF if congestion is detected at some place other than the start (in terms of traffic direction) of the DIF. In this case a backward signalling mechanism is used to relay the signal. Currently only the pure pushback feedback like mechanism is defined in RINA [16].

3.2   Description of the Model

We model Fig. 1 as a fluid type model using Simulink [14]. This allows us to observe the stability dynamics of the system at all levels, especially the uppermost level which is presented here. We use a common generalised congestion control mechanism for the DIFs. Key building blocks in the model are: sender, DIF, and bottleneck. Receivers are modelled only in how they relay congestion signals to the transmission side.

Sender. Figure 2 shows the model used for the sender (End-to-End DIF layer) based on the following differential equation:

$$\dot{r} = \begin{cases} rCK_D & C > 0 \\ (r + I_{\min})K_I & C = 0 \end{cases} \tag{1}$$

where r is the send rate, C the collective measure of congestion ($C = \sum c$, $\forall c$ relevant to the particular feedback scenario), K_I is the increase gain, K_D is the negative decrease gain, and $I_{\min}$ is the minimum increase factor. The sender is

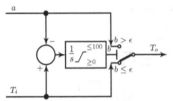

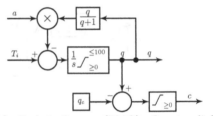

(a) Traffic out Queue (ToutQ). Inputs: proved available rate a and input traffic rate T_i. Output: output traffic rate T_o.

(b) Packet Queue (PktQ). Inputs: link probed available rate a and input traffic rate T_i. Outputs: packet queue size q and local congestion signal c.

Fig. 3. DIF queueing models

greedy, modelling an aggregate of senders, increasing their send rate until they are notified of congestion. The sender's send rate increase is in proportion to its current send rate. This is not the additive increase used in Standard TCP, but more closely resembles the increase mechanism in Scalable TCP [10]. A scalable mechanism makes model more versatile and easier to configure and experiment with, and is also a shortcoming modern TCP enhancements target [2,20]. The response to congestion is proportional to the current send rate and the level of congestion.

When there is no congestion $(C \leq \epsilon)^2$, the rate increases in proportion to the current rate (r) at gain K_I. When there is congestion $(C > \epsilon)$ the rate decreases in proportion to the current rate (r) and the congestion signal $(C \times K_D)$.

Congestion Controled DIFs. The behavior underlying congestion controled DIFs (Access DIF, Network DIF, and Route DIF in Fig. 1) are only different to the sender in that they do not greedily probe for available capacity. Their job is to track the capacity that is available relative to the traffic they must relay. In this way DIFs buffer temporary excesses in traffic and probe for a higher rate when these excesses are sustained. They also reduce their send rate limit when the traffic demand does not require the higher limit. This congestion control behavior is necessary for DIFs and has parallels with recent enhancements of TCP to support application limited traffic [5].

The dynamics of lower DIFs are governed by two queueing models: an instantaneous fluid send queue (ToutQ, see Fig. 3a) that models how traffic is sent through the node, and a fluid approximation of packet queue dynamics (see Fig. 3b) to model congestion.

Traffic out queue. ToutQ (see Fig. 3a) integrates (minimum limited to 0) the difference between the traffic arrival rate (T_i) and the probed available link rate (a). If the queue size is above ϵ, traffic is sent at a rate equal to a. Otherwise traffic is sent at the traffic arrival rate (i.e. $T_o = T_i$).

---

[2] $\epsilon = 0.01$ is used instead of 0 to aid the numerical solvers used in Simulink.

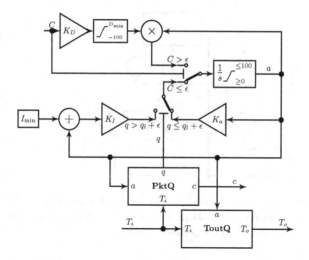

Fig. 4. Congestion control model for DIFs. Inputs: input traffic rate T_i and congestion signal C. Outputs: output traffic rate T_o and local congestion signal c.

Packet queue. The packet queue model (Fig. 3b) is based on an approximation of packet queue dynamics proposed in [1]:

$$\dot{q} = \lambda - \frac{\mu_0 q}{q + 1} \tag{2}$$

where q is the queue length in packets, λ is the traffic arrival rate (T_i in our model), and μ_0 is the maximum service rate (DIF probed available rate, a, in our model). The congestion threshold is set to equilibrium load of about 90 %, which gives rise to an equilibrium packet queue size of 9 packets. The queue size above this threshold (q_c) is fed back as the congestion signal (c) using the various feedback mechanisms being tested.

This M/M/1 based queue is sufficient for the dynamics we wish to analyse. However the same pointwise stationary fluid approximation technique can be applied to more complex arrival and service models as shown in [19], and may be useful in further studies of more complex traffic scenarios.

Complete DIF Congestion Control Model. Figure 4 shows the model used for the DIFs based on the following differential equation:

$$\dot{a} = \begin{cases} aCK_D & C > 0 \\ aK_a & C = 0 \wedge q < q_l \\ (a + I_{\min}) K_I & C = 0 \wedge q \geq q_l \end{cases} \tag{3}$$

were a is the DIF probed available link rate, $C = \sum c$ for all c relevant to the particular feedback scenario, K_I is the increase gain, K_D is the negative decrease gain, $I_{\min}$ is the minimum increase factor, q is the packet queue size, and q_l is a lower packet queue limit. Similar to the sender model (see Fig. 2), when there is

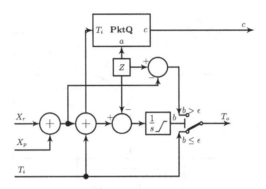

Fig. 5. Bottleneck nodes (x-GW and R-n in Fig. 1). Inputs: random 5% cross traffic X_r, pulse 40% cross traffic X_p (only at R-1 and R-4), input traffic rate T_i. Outputs: output traffic rate T_o and local congestion signal c.

congestion ($C > \epsilon$, where $C = \sum c$ for all c relevant to the particular feedback scenario) a decreases in proportion to its value and the congestion ($C \times K_D$). When there is no congestion and the DIF queue (q) is large, a increases at gain K_I. When there is no congestion and the DIF queue is small, a slowly decreases at gain K_a since a rate of a is not required to carry the traffic presented to the DIF. ToutQ and PktQ are described above and shown in Fig. 3.

Bottleneck Router and Cross Traffic. Bottleneck router nodes do not have any congestion control function, but do produce a congestion signal c. The capacity of the link is $Z = 100$ in these experiments (Fig. 5).

The Bottleneck router nodes have cross traffic X_r and X_p traversing them, restricting the capacity available for T_i to $Z - (X_r + X_p)$. X_r supplies small (average 5% of Z) random perturbations to the available link capacity to model the constantly changing nature of a packet switched network. X_p is used to assess stability in the presence of a large sudden perturbation.

T_o is fed to the next node, with a traffic out queue similar to Fig. 3a used to decide whether T_i can be sent out on the link, or the output rate is the available capacity $Z - (X_r + X_p)$. PktQ is used to measure congestion, see Fig. 3b.

3.3 Connecting the Model Elements

The topology of Fig. 1 is built by linking the traffic and congestion signals of the building blocks. The traffic signals are linked in the model as the information would flow from end-to-end: (i) down the congestion controlled layers of a node, (ii) across the physical link experiencing propagation delay, and (iii) to through the layers[3] up to the uppermost congestion controller in the next node, (iv) repeating from Item i through each node until the receiver.

---

[3] The peer layers on the receiving side do not change the traffic in this model.

That is, the send rate (r) serves as traffic T_i to the Access DIF (see Fig. 1 and Sect. 3.2) whose T_o is delayed and serves as T_i at the bottleneck router S-GW, etc.

The congestion signals, c, are fed back with the relevant hop delays. There is an insignificant delay in the feedback of c within a node. Each congestion control DIF receives $C = \sum c$ appropriate for the particular feedback scenario being modeled.

3.4 Notes on Solving the Model

To aid solving the ordinary differential equations (ODEs) that result from the model, signals are limited in the rate they can change to a gradient of plus and minus twice the Capacity (i.e. 200 in these tests).

4 Simulation Results

To understand system dynamics we look first at how the five different feedback scenarios affect the following key parameters:1. Sender's transmission rate (Tx), 2. Congestion signal (queueing in excess of 90 % load queueing) fed back to the Sender (Cong. FB), 3. The cumulative total of all the queues in the system (TotalQ). We then examine the effect that the gain parameters, K_I and K_D have on the overall stability of the system by looking at the average sender's transmission rate (mean(Tx)) and standard deviation of the sender rate (std(Tx)) for the recursive pushback and recursive backward feedback scenarios.

A large disturbance X_p is introduced at R-1 and R-4 to help assess the stability of the system (see Fig. 1). At $t = [50, 100]$ for R-1 and $t = [150, 200]$ for R-4 the available capacity on the bottleneck is suddenly restricted by 40 % of Z.

4.1 Dynamics of Send Rate, Congestion Signal, and Total Queueing

In these experiments the Sender and all DIFs have identical model gains: $K_D = -0.01$ and $K_I = 0.2$. Figure 6 shows the results for the non-recursive feedback mechanisms as a reference for the recursive feedback mechanisms with both forward and backward congestion signalling. When congestion is closer to the sender (see especially $t = [50, 100]$ s), the benefits of backward congestion notification are most apparent. Note that the end-to-end model used in this scenario has no intermediate DIFs.

We observe that the forward (see Fig. 7a) and backward (see Fig. 7b) recursive feedback congestion control mechanisms perform with similar stability to the non-recursive congestion control. Looking at $t = [0, 10]$ s, notice that the buffers, indicated by TotalQ, begin to fill sooner than they do in the corresponding scenarios in Fig. 6. This is because each DIF tries to keep its load, with respect to the probed available capacity on the link (a), within an acceptable range as measured by the DIF packet queue. Recursive congestion control does have a

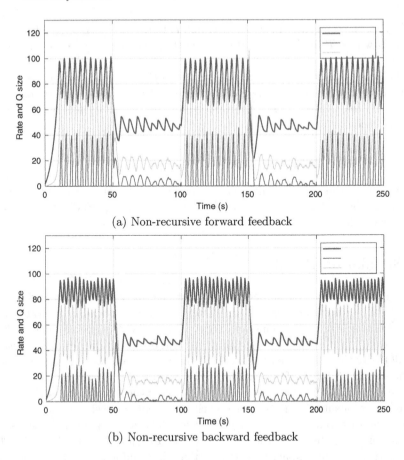

(a) Non-recursive forward feedback

(b) Non-recursive backward feedback

Fig. 6. Non-recursive rich signal feedback. (Tx is the sender transmission rate, Cong. FB is the congestion signal, and TotalQ is the cumulative total of all queues in the system.)

higher total buffer usage. This is not completely unexpected since the recursive architecture has buffers at every DIF. When designed appropriately this buffering can help the network to maintain a high performance in normal fluctuating traffic demands. However, if not designed appropriately they can unnecessarily delay a packet's end-to-end transit.

Using backward feedback (see Figs. 6b and 7b) provides improvements for both the recursive and non recursive feedback mechanisms.

In choosing between the forward and backward congestion notification mechanisms one must consider that the backward mechanism is more difficult to implement and deploy in a layered architecture. Lower layers, working on different aggregates of end-to-end flows, do not have the necessary information to signal the correct sender (in RINA, lower DIFs are not meant to inspect packets for headers from upper DIFs). Each DIF must make a local decision about

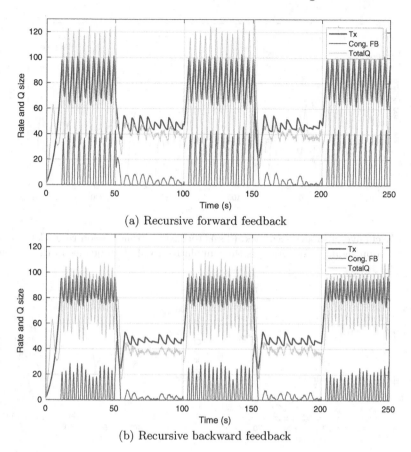

(a) Recursive forward feedback

(b) Recursive backward feedback

Fig. 7. Recursive rich signal feedback. (Tx is the sender transmission rate, Cong. FB is the congestion signal, and TotalQ is the cumulative total of all queues in the system.)

which of the sources of incoming traffic should be informed of congestion, and to what extent. This decision will influence the fairness of rate allocations. Forward signalling is easier to implement as it can follow the path of the packet up the layers to be reflected by receivers as necessary.

The pushback mechanism has the nice architectural property of keeping the congestion control at each DIF layer independent. Unfortunately, this results in poor performance (see Fig. 8) with continuous large oscillations in send rates. To investigate the reasons for this we look in detail at the DIF buffer dynamics of the recursive backward feedback and pure pushback mechanisms at the $t = 50\,$s cross traffic perturbation at ① in Fig. 1.

Looking first at the recursive backward feedback queues shown in Fig. 9a, when the sudden increase in cross traffic starts at $t = 50\,$s, the queue at the affected router begins to grow ($R\text{-}1$). This causes $Rt.\ DIF$ and $Nt.\ DIF$ operating across that path to receive a congestion signal and reduce its allowable sending

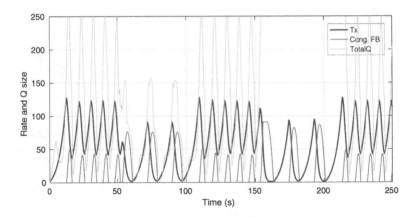

Fig. 8. Recursive pushback feedback. (Tx is the sender transmission rate, Cong. FB is the congestion signal, and TotalQ is the cumulative total of all queues in the system.)

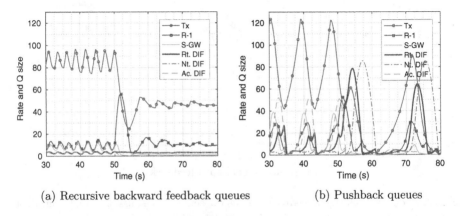

(a) Recursive backward feedback queues   (b) Pushback queues

Fig. 9. Comparing recursive pushback and backward feedback queue growth. Referring to Fig. 1: Tx→send rate; queue sizes at: R-1→router where congestion is induced (Router DIF), S-GW→gateway router closest to the sender, $Rt.$ DIF→Route DIF at ISP 1 access, $Nt.$ DIF→Network DIF at ISP1 Access, $Ac.$ DIF→Access DIF at sender.

rate (a in Fig. 4). There is no delay between the reaction of $Rt.$ DIF and $Nt.$ DIF as they are in the same network element. Due to the start of congestion just before $t = 50$ s, the sender had already started reducing its rate so the queues at $Rt.$ DIF do not increase. The reduction in traffic alleviates congestion at R-1, though it remains the bottleneck. In this experiment all DIF congestion signal gains (K_D) and increase gains (K_I) are the same. Some performance improvement may be able to be made by tuning K_D and K_I by taking into account the collection of congestion signals that collectively make up the C input to each DIF, and round trip time a particular DIF operates over. However, this is beyond the scope of this paper and an area of future work.

The interactions in the pure pushback control scenario shown in Fig. 9b are more complex. There is general instability caused by the inter-DIF interactions. However, looking just at the disturbance at $t = 50$ s, we see that this occurs when the sender is at one of the lower points in its large send rate (Tx) oscillations. The queue at $R\text{-}1$ begins to grow, which causes $Rt.$ DIF to reduce its rate, causing $Rt.$ DIF's queue to grow and eventually $R\text{-}1$'s queue to shrink. $Nt.$ DIF also reduces its a, but $Nt.$ DIF's queue does not grow immediately as might be expected since the traffic arriving at $Nt.$ DIF was initially less than a due to the previous rate cycling (note the capacity of the bottleneck is between 50 and 55). As Tx increases and $Nt.$ DIF's a decreases, $Nt.$ DIF's queue rapidly grows; $Rt.$ DIF's queue shrinks as $Nt.$ DIF restricts traffic flow. $Nt.$ DIF's queue growth signals the sender to reduce its rate (Tx).

Notice the push-back queue behaviour caused by the DIF interaction from $t = 65$ s, $R\text{-}1$ push-back to $Rt.$ DIF, push-back to $Nt.$ DIF. This is repeated until congestion due to the large cross traffic disturbance stops at $t = 100$ s with the sender rate going from 0 to a peak and back to 0 again. The system is not able to obtain an efficient stable state. There are two key reasons for the instability in the pure pushback mechanism: 1. It takes time for the queueing signal to be pushed back, 2. The sender is not aware of the total queueing along the path—congestion at lower DIFs is not feedback to the sender.

4.2 Effect of DIF Gains on Stability

Since DIFs under the end-to-end Sender DIF operate over a smaller topology, there may be some advantage in their gains being more aggressive than the Sender's. Leaving K_I and K_D the same for the Sender, we adjust the lower DIF gains, multiplying them by the factor $g = 0.5, 1, \ldots, 5$. Figure 10 illustrates the effect that varying K_I and K_D in the lower DIFs has on the overall system stability for the recursive backward feedback (RB) and the pushback (PB) scenarios. The simulation results have been re-sampled to avoid bias in points calculated through Simulink's variable step solver. Mean(Tx) is the average over the entire 250 s, while std(Tx) is calculated using the variance about the mean of each 50 s interval, matching the step changes in available rate in the simulations. The recursive pure pushback mechanism suffers from poorer throughput and a much higher variance than the recursive backward feedback mechanism. The performance of both mechanisms generally diminishes as the lower DIF congestion controls become more aggressive than the Sender ($g \geq 1$), though to a lesser extent with the recursive backward feedback mechanism. When the lower DIFs are less aggressive than the sender the recursive backward feedback mechanisms has a small drop in performance as the senders rate adjustments are limited by the DIFs. This dampening helps reduce the standard deviation of Tx a little in the pushback scenario.

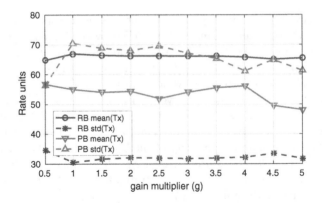

Fig. 10. Effect of increase and decrease gain ($g \times K_I$ and $g \times K_D$ for all lower DIFs) on mean and standard deviation of theresulting sender transmit rate (Tx).

4.3 Observations from Conducting Experiments

Constructing the model and conducting these experiments gave insight into the dynamics of congestion control in recursive architectures. In this section we outline some of these insights.

Congestion control with recursive feedback may require more buffering between end points than end-to-end congestion control. This is particularly apparent in these experiments where all physical links have the same capacity, with the available capacity only being slightly different due the cross traffic. Whether this significantly affects end-to-end latency needs further investigation.

For the recursive congestion controls to work well each DIF needs to accurately track the available capacity, which happens in these experiments. This is especially critical for pushback. A DIF only probes for more available capacity when its buffers begin to fill, and reduces its maximum send rate when its buffers reduce or it is notified of congestion. In the topology we model where all links have a similar available capacity, this can result in cascading delays as each DIF probes for more capacity to meet demand. In the pure pushback scenario this probing and corresponding reactions to congestion result in wild oscillations.

These experiments suggest that the pushback mechanism is the least stable of the five feedback mechanisms tested. We explored manually tuning the pushback mechanism's parameters to yield better results, but with little success. In experiments with a simpler topology, half the size and with one less layer, we were able to manually tune the pure pushback to produce some improvement. Even so, in the smaller simpler topology, it was still the least stable and worst performing of the five feedback mechanisms.

Since the sender is the ultimate source of traffic, it is important that it receives a complete picture of congestion along the end-to-end path in order to react appropriately to it. The other two recursive congestion control feedback mechanisms give DIFs and the sender a complete picture of congestion along their respective end-to-end controlled paths, and performed equivalently well as a non-recursive mechanism. They therefore seem to be the best way forward for a stable recursive congestion control that will not diminish the advantages of using a recursive network architecture such as RINA and RNA.

These experiments use identical parameters for all congestion control DIFs in Sect. 4.1. This makes it easier to compare mechanisms, but is not an optimal solution for a recursive architecture where each DIF may be able to adapt to the topology beneath it and use their observation of feedback delays and optimise their responses accordingly. Section 4.2 shows that simply making the congestion control in lower DIFs more aggressive does not give performance benefits. The experiments do not show how well an individually tailored and optimised recursive congestion control for each DIF may be able to work; an area for further investigation. To date such a congestion control mechanism does not exist, and the interactions demonstrated in this paper indicate that a stable efficient design may be difficult. Backstepping control theory [12] has similarities with recursive architecture congestion control and may hold promise in designing such a mechanism.

5   Conclusions and Future Work

Our findings demonstrate that the layer independent pure pushback mechanism performs badly in a realistic topology. Delays in the congestion signal being pushed back and up through the layers are observed to be part of the problem. Also we observe that due to the strict layer independence, congestion controllers at each layer—and ultimately the sending sources—are unable to respond to the full extent of congestion since it is hidden from them.

We observe that a feedback mechanism which allows all congestion controllers to receive signals from all sources of congestion along the path they control allows for a better performing and more stable control. Backward feedback provides the greatest benefits, especially for sources of congestion that are closer to the sending sources than the destinations. However, we acknowledge that this type of feedback can be difficult to implement in a real recursive architecture. Forward end reflected feedback does not perform quite as well, but does still provide a workable mechanism.

Work on congestion control in recursive architectures is in its infancy. The area itself is very large, and this work takes the first steps at developing an understanding for the way feedback mechanisms affect the recursive congestion control system in a simple but non-trivial network. We plan to continue work in this area investigating the use of multiple loop control theory, such as backstepping control, as a means of developing a stable efficient mechanism for congestion control in recursive network architectures such as RINA and RNA.

Acknowledgment. This work has received funding from the European Union's FP7 research and innovation programme under grant agreement No. 619305 (PRISTINE). The views expressed are solely those of the authors.

References

1. Agnew, C.E.: Dynamic modeling and control of congestion-prone systems. Oper. Res. **24**(3), 400–419 (1976)
2. Bensley, S., Eggert, L., Thaler, D., Balasubramanian, P., Judd, G.: Datacenter TCP (DCTCP): TCP Congestion Control for Datacenters. Internet-Draft draft-ietf-tcpm-dctcp-01, Internet Engineering Task Force, May 2016. https://tools.ietf.org/html/draft-ietf-tcpm-dctcp-01. work in Progress
3. Day, J.: Patterns in Network Architecture: A Return to Fundamentals. Prentice Hall, Upper Saddle River (2007)
4. Day, J., Matta, I., Mattar, K.: Networking is IPC: a guiding principle to a better internet. In: Proceedings of the ACM CoNEXT, p. 67 (2008)
5. Fairhurst, G., Sathiaseelan, A., Secchi, R.: Updating TCP to support rate-limited traffic. RFC 7661, RFC Editor, October 2015
6. Gerla, M., Kleinrock, L.: Flow control: A comparative survey. IEEE Trans. Commun. **COM–28**(4), 553–574 (1980). Also published in Computer Network Architectures and Protocols. Greed, P. (ed.) Plenum Press, pp. 361–412 (1982)
7. Gursun, G., Matta, I., Mattar, K.: On the performance and robustness of managing reliable transport connections. Technical report, CS Department, Boston University, bUCS-TR-2009-014 (2009)
8. Hassan, M., Jain, R.: High Performance TCP/IP Networking - Concepts, Issues, and Solutions. Pearson Education, Upper Saddle River (2004)
9. Ishakian, V., Akinwumi, J., Esposito, F., Matta, I.: On supporting mobility and multihoming in recursive internet architectures. Comput. Commun. **35**(13), 1561–1573 (2012)
10. Kelly, T.: Scalable TCP: improving performance in highspeed wide area networks. ACM SIGCOMM Comput. Commun. Rev. **33**(2), 83–91 (2003)
11. Khalil, H.K.: Nonlinear Systems. Prentice Hall, Upper Saddle River (2001)
12. Krstic, M., Kokotovic, P.V., Kanellakopoulos, I.: Nonlinear and Adaptive Control Design, 1st edn. Wiley, New York (1995)
13. Lopez, D. (ed.): Use cases description and requirements analysis report. PRISTINE project, May 2014. http://ict-pristine.eu/wp-content/uploads/2013/12/pristine-d21-usecases_and_requirements_v1_0.pdf
14. The Mathworks, Inc., Natick, Massachusetts: MATLAB SIMULINK version 8.7 (R2016a) (2016)
15. Small, J.: Patterns in network security: an analysis of architectural complexity in securing recursive inter-network architecture networks. Master's thesis, Boston University Metropolitan College (2012)
16. Teymoori, P., Welzl, M., Stein, G., Grasa, E., Riggio, R., Rausch, K., Siracuss, D.: Congestion control in the recursive internetwork architecture (RINA). In: IEEE International Conference on Communications (ICC), Next Generation Networking and Internet Symposium, May 2016
17. Touch, J., Baldine, I., Dutta, R., Finn, G.G., Ford, B., Jordan, S., Massey, D., Matta, A., Papadopoulos, C., Reiher, P., Rouskas, G.: A dynamic recursive unified internet design (DRUID). Comput. Netw. **55**(4), 919–935 (2011). http://www.sciencedirect.com/science/article/pii/S138912861000383X. Special Issue on Architectures and Protocols for the Future Internet

18. Touch, J.D., Pingali, V.K.: The RNA metaprotocol. In: Proceedings of 17th International Conference on Computer Communications and Networks, ICCCN 2008, pp. 1–6. IEEE (2008)
19. Wang, W.P., Tipper, D., Banerjee, S.: A simple approximation for modeling nonstationary queues. In: Proceedings of the IEEE International Conference on Computer Communications (INFOCOM), vol. 1, pp. 255–262, March 1996
20. Welzl, M.: Network Congestion Control: Managing Internet Traffic (Wiley Series on Communications Networking & Distributed Systems). Wiley, Hoboken (2005)

A PEPA Model of IEEE 802.11b/g with Hidden Nodes

Choman Othman Abdullah[1,2(✉)] and Nigel Thomas[2]

[1] School of Science Education, University of Sulaimani, Sulaymaniyah, Iraq
`choman.abdullah@univsul.edu.iq, c.o.a.abdullah@ncl.ac.uk`
[2] School of Computing Science, Newcastle University, Newcastle upon Tyne, UK
`nigel.thomas@ncl.ac.uk`

Abstract. The hidden node problem is a well known phenomenon in wireless networks. It occurs when two nodes transmit which are out of range of each other, but both within range of at least one of the intended recipients. This results in a non-delivery which is generally only detectable by the sender due to a lack of acknowledgement. In this paper we explore the performance of IEEE 802.11 b and g subject to hidden nodes using the stochastic process algebra PEPA. We show that faster transmission yields better maximum throughput and the slower the speed of transmission relative to the inter-frame spacing (IFS) duration, the greater the probability of collision in transmission.

Keywords: IEEE 802.11b/g · Performance modelling · PEPA · Hidden node

1 Introduction

The IEEE 802.11 family of protocols has become the standard for wireless local area networks [1,13]. The different protocols within 802.11 (a/b/g/n/ac) all have a similar structure, but are defined to work over different ranges and at different transmission rates. For example, IEEE 802.11b operates at up to 11Mb/s in 2.4 GHz, while 802.11 g enhances the data rates up to 54 Mb/s within the same bands [21]. Clearly, a greater transmission speed should result in greater throughput. However, there are topological effects which mean a given network might not be able to maintain the optimal performance for all nodes. For example, in our previous work [2,4], we considered a situation where a node attempting to transmit might always be out competed by its neighbours, leading to an unfair sharing of network bandwidth. In this paper we consider another topological effect which affects performance, the hidden node problem.

The hidden node problem is well known in wireless networks. It arises when two nodes attempt to transmit which are out of range of one another (and hence cannot detect each other's transmission) but one or both of the intended recipients is within range of both transmitting nodes. Thus the recipient will only hear the distorted signal created by the interference of the overlaid transmissions and cannot therefore receive the its intended message. In the general case,

© Springer International Publishing AG 2016
D. Fiems et al. (Eds.): EPEW 2016, LNCS 9951, pp. 126–140, 2016.
DOI: 10.1007/978-3-319-46433-6_9

the transmitting nodes will not be able to detect this interference and so will not know that there has been a collision. In some protocols the receiving node might transmit a jamming signal, which would have the effect of resetting any transmissions. However, it is more likely that the transmitting nodes will only know that their message was unsuccessful because they will not receive an acknowledgement from the recipient. They will then attempt to resend the failed message, with possibly the same outcome. It should be clear that there is no simple way to avoid the hidden node problem and that it may have a significant effect on network performance. As such modelling situations with hidden nodes is clearly of practical interest.

In this paper we explore a model of the hidden node problem in IEEE 802.11 b and g where access is controlled by the Distributed Coordination Function (DCF). The DCF is the de-facto standard at the MAC (Medium Access Control) layer of IEEE 802.11. IEEE 802.11 b and g use CSMA/CA (Carrier Sense Multiple Access/Collision Avoidance) to try to minimise the occurrence of collisions between simultaneously transmitted data. However, CSMA/CA is only effective when nodes can detect other transmitting nodes, which is not the case if a competing node is out of transmission range. Hence, in our model no signal will be detected before transmission and so if the other node is already transmitting then a collision will definitely occur. Once the collision is detected (through the lack of an acknowledgement) then the node will enter its backoff process in an attempt to avoid a repeated of the collision.

Performance modelling has been employed successfully to evaluate the performance of (current and future) networking systems for many decades. There have been many attempts to model different aspects of 802.11 using a wide variety of methods. The majority of these studies have used simulation, which can give a good indication of predicted performance, but provide limited insight on the behaviour which leads to this performance. Formal modelling techniques, such as stochastic process algebra, allow the modeller to reason about properties of a model via explicit naming of components and actions, but constructing large process algebra models with complicated internal component behaviours is a difficult task. As a result there are only a few of published studies which have used process algebra to model aspects of 802.11 [2,4,7,18]. Our model is defined using the stochastic process algebra PEPA [14] based on an existing model of IEEE 802.11b by Kloul and Valois [18]. We extend the previous work by also considering IEEE 802.11g, which uses the same method but differs in its transmission rates and inter-frame spacing (the delays incurred between actions). We further compare the results obtained from 802.11 g with those from 802.11 b and show some interesting similarities in performance profiles.

The structure of this paper is organised as follows: Sect. 2 includes a background and related work to give an overview of IEEE 802.11 and PEPA. The model and basic access mechanism that we used in PEPA for our model is shown in Sect. 3. The parameters that we have used are shown in Sect. 4. The results and figures are discussed in Sect. 5 for both 802.11b and g protocols. Finally, conclusion and future works are provided in Sect. 6.

2   Background and Related Work

2.1   IEEE 802.11

The availability of wireless local area networks (WLANs) has increased dramatically over recent years due to the advantage of low installation cost, easy sharing and increasingly high speed. The 802.11 protocols have been deployed widely in wireless devices and have been used commonly as a basic standard for WLANs [1]. The different protocols (a/b/g/n/ac) all have a similar structure, but different operating ranges (power, data rate, frame length etc.) [17]. As the data rate increases these protocols employ increasingly more sophisticated mechanisms. As a consequence of the proliferation of these protocols, there have been many performance studies considering different properties and issues [6,22].

IEEE 802.11g was introduced in 2003 as a compatible extension to IEEE 802.11b over the 2.4 GHz frequency [23]. Vucinic et al. in [25] considered the performance degradation in 802.11g in terms of access delay for dissimilar nodes and throughput, as they analysed collision probability, channel access delay and throughput. Kuptsov et al. assessed fairness in 802.11g by studying the backoff and contention window mechanisms [20]. Here poor fairness arises as unsuccessful nodes are obliged to remain unsuccessful in term of channel access, while the standard backoff protocol allows successful nodes to access the medium successfully for long periods. In our previous work [2,4] we analysed the (un)fairness of 802.11b/g due to pathologic topological effects, but we did not consider the hidden nodes scenario.

A small number of analytical studies have been proposed considering the effect of the hidden nodes on the performance of IEEE 802.11. An analytical model has been presented in [26] to derive the saturation throughput of MAC protocols based on RTS/CTS method in multi-hop networks under the assumption of heavy traffic load. In [15] the throughput of the IEEE 802.11 DCF scheme with hidden nodes in a multi-hop ad hoc network was analysed when the carrier sensing range is equal to the transmission range. Hou et al. [16] undertook an analytical study to derive the throughput of IEEE 802.11 DCF with hidden nodes in a multi-hop ad hoc network. The main drawback of this work is that the state of retransmission counter is not taken into account when obtaining the collision probability. A simple analytical model has been presented in [26] to derive the saturation throughput of MAC protocols based on RTS/CTS method in multi-hop networks. The model was only validated under heavy traffic assumption. Ekici and Yongacoglu [9] proposed an analytical model for IEEE 802.11 DCF in symmetric networks in the presence of the hidden nodes and unsaturated traffic. The model assumes that the collision probability is constant regardless of the state retransmission counter. Younes and Thomas [29] presented an SRN model of IEEE 802.11 with hidden nodes and multiple hops. One advantage of a formal model such as this is that different protocols can be compared and models can be adapted as new versions of the protocol are developed.

Slowly, IEEE 802.11n is replacing the old protocols, although it still coexists with others, such as IEEE 802.11g. Although this paper only considers IEEE

802.11b and g, the results are still of interest in IEEE 802.11n. Galloway [11] has studied on the effects of coexisting both 802.11n and 11g in wireless devices. IEEE 802.11n is MIMO "Multiple Input, Multiple Output" antenna provides higher speed, wide range and reliability over IEEE 802.11b/g. Many researchers have studied IEEE 802.11n in terms of PHY values to increase the higher data rates and MAC enhancements to reduce overhead via various aspects such as single with multiple rates and ACK with delay ACK [10, 28].

2.2 PEPA

Performance Evaluation Process Algebra (PEPA) [14] is a process algebra which provides a useful modelling formalism to investigate properties of protocols and other well defined systems like multimedia applications and communication systems. PEPA models are specified in terms of components which interact through shared actions. In PEPA, actions have a duration which is determined by a rate parameter of the negative exponential distribution. It is shared actions, where a rate may be given by one or both interacting components and the shared rate is determined by the slowest participant. In network protocols, components can be network nodes and the transmission media and shared actions can be thought of as the transmission of messages (packets) from one node to another through the medium. The combination of all components into a single system gives rise to labelled transition system where the transitions between states are negative exponentially distributed actions, hence the resultant system is a continuous time Markov chain (CTMC). The PEPA Eclipse Plug-in tool [12] supports a range of powerful analysis techniques for Markov Process (CTMC), systems of ordinary differential equations (ODE) and stochastic simulation which allows modellers to derive results (both transient and steady state) with relative ease.

Despite the benefits of PEPA, there are few examples in the literature where it has been used to study IEEE 802.11 especially the 802.11n. Argent-Katwala *et al.* [7] studied WLAN protocols and performance models of the 802.11 in terms of its QoS based on PEPA. They argued that most of the technologies have been developed to enhance the reliability of computer networks. In wireless communication protocols security is mandated in exchanging data, which must be delivered within a specific time. They used PEPA to find properties which cannot be easy to find manually in terms of computing quantitative, passage time and increase higher probability for performance demands. Sridhar and Ciobanu [24] used PEPA and π-calculus to study DCP within IEEE 802.11, which uses (CSMA/CA) and backoff mechanism. They analysed the handoff mechanism and channel mobility. Kloul and Valois [18] developed two models of network topologies which an effect on the performance of IEEE 802.11b. In one scenario they considered unfairness caused by competing neighbours and in the other scenario they considered the hidden node problem. They validated their results using simulation. Abdullah and Thomas [2] extended the analysis of neighbourhood competition in 802.11b and then extended it to 802.11g [4]. More recently they have considered the effect of variable frame transmission duration on fairness [3]. In this paper we extend the hidden nodes model of Kloul and Valois [18] to consider 802.11g and compare results against 802.11b.

3   The Model

3.1   Basic Access Mechanism

The Basic Access (BA) method is widely used with the IEEE 802.11 protocols. Fundamentally, it cooperates in one of two different modes. The first mode is Point Coordination Function (PCF) and the second mode is Distributed Coordination Function (DCF). PCF needs a central control object and DCF is based on CSMA/CA. The basic access mechanism in 802.11b and g is DCF, which is a common technique used up to 802.11g [5]. The DCF mechanism specifies two techniques for data transmission, which are the basic access method and two way handshake mechanism, in this study we focused on the basic access method. In a WLAN, a node senses the medium to discover if it is free to use; if so, then the node can make its transmission. On successful receipt, a receiving node will transmit an acknowledgement (ACK). However, if two nodes within transmission range attempt to transmit simultaneously, then collision occurs resulting in an unsuccessful transmission and an initiation of the backoff algorithm. An unsuccessful node waits for a random time (backoff) in the range $[0, CW]$, where the contention window CW is based on the number of transmission failures. The initial value of CW is 31 for 11b and 15 for 11g and it is doubled after every unsuccessful transmission, until it reaches to the maximum number 1023 (see [8,17] for detailed explorations of the backoff algorithm). CW returns to the initial value after each ACK received. When the backoff period has expired, the node again senses the network to see if it is free to use. The aim of the backoff is to try to avoid repeated collisions between competing nodes, as it is unlikely that two nodes will choose the same random backoff period. The more collisions occur, the larger the contention window and hence the less likely that another collision will occur. If the medium is sensed to be busy then the node will wait for a period before retrying. This is so that multiple waiting nodes will not immediately try to transmit once the medium is quiet, which would obviously cause a collision (see [19]).

If all nodes can hear all other nodes, i.e. they are all within sensing range of each other, then the BA method will eliminate almost all collisions. There would still be a very small window when collisions could occur, which would be the time it takes a signal to traverse the sensing range, but this would be relatively insignificant in such small high speed networks. However, in practice most networks cover a much larger area than the sensing range of a single node. Therefore there is a possibility that two nodes which lie outside each other's sensing range will choose to transmit simultaneously to nodes which are within the transmission range of both senders. Thus, although the senders cannot hear each others transmission, an intended recipient will hear both transmissions overlaid. This results in interference and hence the non-delivery of the frame. Such a frame would clearly not be successfully received and so an acknowledgement would not

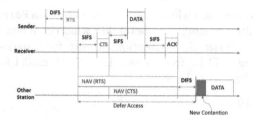
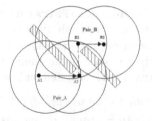

Fig. 1. Basic access mechanism   **Fig. 2.** Hidden node

Table 1. Attribute values of IEEE 802.11b/g

Attribute	Typical value	
	802.11b	802.11g
CWmin, and CWmax	31, and 1023	15(pure), and 1023
Slot time	20 µs	20 µs, 9 µs
SIFS	10 µs	10 µs
DIFS	50 µs	50 µs, 28 µs
EIFS	364 µs	364 µs

be sent. The sending nodes would wait for an acknowledgement for a period and then start the backoff process once determining that the acknowledgement is not forthcoming. This scenario characterises the hidden node problem. Clearly, each transmitting node cannot sense the other and therefore such collisions are inevitable and consequently bandwidth is wasted and there is an impact on performance.

3.2 Scenario with a PEPA Model

In this section we present the model of hidden nodes in the 802.11b/g, as illustrated in Fig. 2, by using PEPA. The PEPA model is similar for both protocols with some different parameters and they are consistently attempts to occupy the medium. We used the same model to study for 802.11b and 802.11g protocols.

Hidden Nodes Scenario. In this study the hidden nodes scenario is modelled as two communicating pairs of nodes, `PairA` and `PairB`, which interact over a transmission medium `MediumS`. This scenario is not free of collision and it happens when each pair attempts to transmit simultaneously (as they cannot sense each other). While the first node is "listening" on the network it can access the channel as it is free to send any packets, meanwhile, the second one cannot sense the medium is occupied by the first one as its hidden. `PairA1` and `PairB1`, `PairA1` and `PairB2` and `PairB2` and `PairA2` are independent respectively, see Fig. 2. When `PairA` attempts to transmit and access the medium, the `PairB` is hidden from it, so `PairA` cannot receive acknowledgement, in this situation `PairA` starts to retransmit the packets after resizing the contention window. When a collision occurs and `PairA` waits or resize the contention window for a period of time in case if the channel is busy till its free to use. The PEPA model for `PairA`, `PairB` and `MediumS` is specified as follows:

$$PairA \stackrel{def}{=} (draw_b\, ackoff, r).PairA0$$
$$PairA0 \stackrel{def}{=} (count_d\, ifs, mudifs).PairA1$$
$$PairA1 \stackrel{def}{=} (count_b\, ackoff, pmubck).PairA1 + (end_b\, ackoff, qmubck).PairA2$$

$$PairA2 \stackrel{def}{=} (transmit, mudata).PairA3$$
$$PairA3 \stackrel{def}{=} (count_s\, ifs, musifs).PairA6$$
$$PairA6 \stackrel{def}{=} (ack, muack).PairA + (collision, \top).(resize_W, s).PairA$$
$$PairB \stackrel{def}{=} (draw_b\, ackoff, r).PairB0$$
$$PairB0 \stackrel{def}{=} (count_d\, ifsB, mudifs).PairB1$$
$$PairB1 \stackrel{def}{=} (count_b\, ackoffB, pmubck).PairB1 + (end_b\, ackoffB, qmubck).PairB2$$
$$PairB2 \stackrel{def}{=} (transmitB, mudata).PairB3$$
$$PairB3 \stackrel{def}{=} (count_s\, ifs, musifs).PairB6$$
$$PairB6 \stackrel{def}{=} (ackB, muack).PairB + (collision, \top).(resize_W, s).PairB$$

$$MediumS \stackrel{def}{=} (transmit, \top).MediumS1 + (transmitB, \top).MediumS2$$
$$MediumS1 \stackrel{def}{=} (ack, \top).MediumS + (transmitB, \top).MediumS3$$
$$MediumS2 \stackrel{def}{=} (transmit, \top).MediumS3 + (ackB, \top).MediumS$$
$$MediumS3 \stackrel{def}{=} (collision, rc).MediumS$$

The complete system: In this PEPA model all components are interacting with this cooperation sets:

$$Set \stackrel{def}{=} ((PairA) \underset{\mathcal{K}}{\bowtie} PairB) \underset{\mathcal{L}}{\bowtie} MediumS$$

where $\mathcal{K} = \{collision\}$ and $\mathcal{L} = \{transmit, ack, transmitB, ackB, collision\}$
where $\mathcal{K} = \{collision\}$ and $\mathcal{L} = \{transmit, ack, transmitB, ackB, collision\}$.

4 Parameters

Inter-frame spacing is very specific in the IEEE 802.11, as it coordinates access to the medium to transmit frames. For convenience, each pair in this model has count backoff and end backoff actions with rates $(p \times \mu bck)$ and $(q \times \mu bck)$ respectively and we assume the values of p and q $(q = 1 - p)$ are equal to 0.5. According to the 802.11b and g definition, the data rate per stream are (1, 2, 5.5, and 11) Mbits/s and (6, 9, 12, 18, 24, 36, 48, and 54) Mbits/s (see [8,17] for more details). In this paper we considered 6, 12, 36 and 54 Mbits/s as a sample of data rates for 802.11g, these rates have been applied with each of the packet payload size (700, 900, 1000, 1200, 1400 and 1500) bytes. The packets per time unit for arrival and departure rate are $\lambda oc = 100000$ and $\mu = 200000$ respectively. In this model (μack) shows as a rate of ACK of packages, where $\mu ack =$ Channel throughput $\div$ (Ack length $= 1$ byte).

4.1   Inter-Frame Space (IFS)

In 802.11 before each frame transmits, the length of the IFS depends on the previous frame type, if noise occurs, the required IFS is used. Possibly, if transmission of a particular frame ends and before another one starts the IFS applies a delay for the channel to stay clear. It is an essential idle period of time needed to ensure that other nodes may access the channel. The purpose of an IFS is to supply a waiting time for each frame transmission in a particular node, to allows the transmitted signal to reach another node (essential for listening). IEEE 802.11 have several IFS: SIFS, DIFS, EIFS and Slot time, see [8,27].

Short Inter-Frame Space (SIFS). SIFS is shortest IFS for highest priority transmissions used with DCF, measured by microseconds. It is important in 802.11 to better process a received frame. SIFS $= 10\,\mu$s in 802.11b/g/n.

DCF Inter-Frame Space (DIFS). DIFS is a medium priority waiting time after SIFS and mcuh longer to monitor the medium. If the channel is idle again, the node waits for the DIFS. After the node determines that the channel is idle for a specific of time (DIFS) then it waits for another (backoff).

$$\text{DIFS} = \text{SIFS} + (2 \times (\text{Slot time} = 20\,\mu\text{s in 802.11b/g/n})).$$

Extended Inter-Frame Space (EIFS). When the node can detect a signal and DIFS is not functioning during collision, the transmission node uses EIFS instead of DIFS, (used with erroneous frame transmission). It is the longest of IFS but has the lowest priority after DIFS. in DCF it can derive by:

$$\text{EIFS} = \text{SIFS} + \text{DIFS} + \text{transmission time}(\text{Ack} - \text{lowest basic rate}).$$

4.2   Contention Window (CW)

A node waits to minimise any collision once experiments an idle channel with appropriate IFS (otherwise many waiting nodes might transmit simultaneously). In CSMA/CA, before sending any frame the node waits a random time backoff, it is selected by node from a Contention Window (CW). A Node needs less waiting time if there is a short backoff period, so transmission will be faster too, unless there is a collision. Backoff is chosen in $[0, CW]$. $CW = CW min$ for all nodes if a node successfully transmits a packet, then receives an ACK. In the non-transmission case, the node deals another backoff, with each unsuccessful transmission it increments by multiplication of 2 at every retransmission for the same packet, this attempt and CW increases exponentially until it reaches $CW max$. Finally, the CW resets to $CW min$ when the packet is received properly. $CW min = 31$ (802.11b), 15 (802.11g) and $CW max = 1023$ (for both 802.11b/g). $CW min$ augmented by 2n-1 on each retry.
Backoff Time $=$ (Random () mod ($CW + 1$)) $\times$ Slot Time.

If BackoffTimer $= b$, where b is a random integer, also $CW\,min \le b \le CW\,max$
The mean of CW is calculated by: μbck $= 10^6 \div$ (Mean of $CW \times$ Time Slot).
The mean of $\mu bck = 15$ (for 802.11b), 7.5 (for 802.11g) and Time slot $= 20\,\mu s$.
The receiver sends an ACK if it gets a packet successfully, it is a precaution
action to notify when collisions occur.

4.3 Data Rates

ACK send by receiver when it gets the packet successfully, it is precautions
action when collisions occur. ACK in 802.11b protocol is deal with data rate (1,
2, 5.5. and 11) Mbits/s, each μack is equal to (1644.74, 3289.5, 9046.125 and
18092.25) Bytes/s respectively. Then, the value of $\mu data$ can be obtained by
(Data rate $\times$ ($10^6 \div 8$)) $\div$ Packet payload size. Similarly, ACK in 802.11g deals
with data rate (6, 12, 36 and 54) Mbits/s and each μack is equal to (9868.44,
19736.88, 59210.64 and 88815.96) respectively.

5   Results of Hidden Nodes Scenario in 802.11b/g

We now use the model presented above to measure the utilisation, probability of
transmission, throughput and collision probability. The channel utilisation rate
for both pairs (A and B) is found by:

$$P[\text{MediumS} \wedge (\text{A2}||\text{B2})] + P[\text{MediumS1}] + P[\text{MediumS2}] + P[\text{MediumS3}].$$

Figure 3 shows the channel utilisation for 802.11b. We can see that for slow
transmission speeds the channel is almost completely saturated, but for faster
transmission there is a fair amount of unused capacity. This is because the inter-
frame spaces are fixed for all transmission speeds and they have to be long enough
to cope with the slowest transmission rate. Therefore in our model, which aims

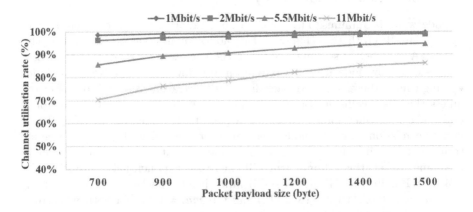

Fig. 3. Channel utilisation rate for both pairs in (802.11b)

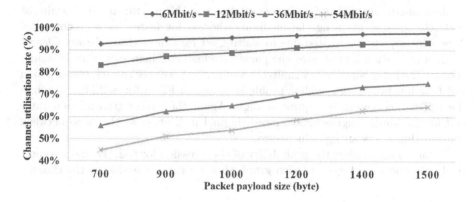

Fig. 4. Channel utilization rate for both pairs in (802.11g)

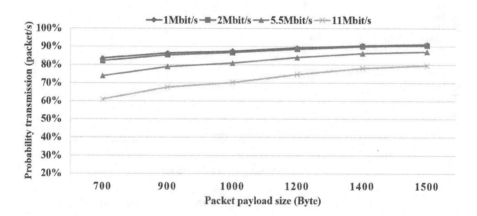

Fig. 5. Probability transmission of (A and B) in (802.11b)

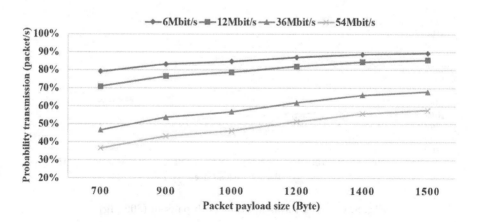

Fig. 6. Probability transmission of (A and B) in (802.11g)

to show maximum utilisation for two nodes, the 1Mbps transmission is almost perfectly efficient at using the medium, whereas for faster transmission rates some capacity will be wasted due to waiting set for slower transmission. Clearly channel utilisation increases as the packet payload size increases. This is simply because the ratio between transmitting and waiting reduces as each transmission will take longer. A very similar profile is shown in Fig. 4 for 802.11g, although the utilisation here is not quite as high. Again the slower transmission rates and longer frame lengths create more channel utilisation as the ratio between transmitting and waiting is increased.

Figures 5 and 6 show the probability of transmission for 802.11b and g respectively. As one would expect, each graph shows a similar profile to the channel

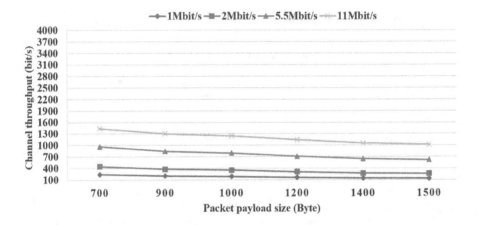

Fig. 7. Channel throughput for both pairs in (802.11b)

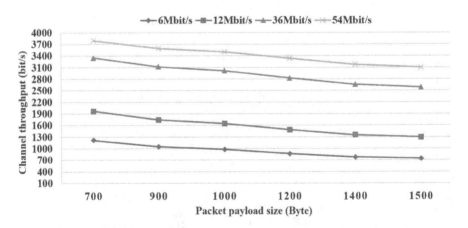

Fig. 8. Channel throughput for both pairs in (802.11g)

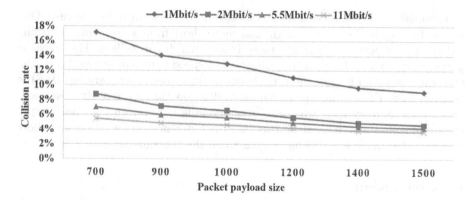

Fig. 9. Probability of collision of transmission in (802.11b)

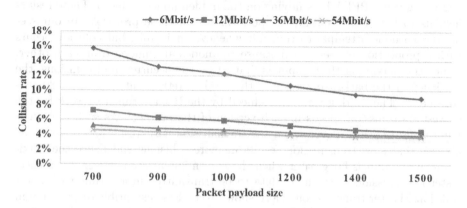

Fig. 10. Probability of collision of transmission in (802.11g)

utilisation, but slightly reduced. What is slightly surprising here is that for the fastest shorted frames in 802.11g, only around 36 % of capacity is being used successfully.

Channel throughput is shown in Figs. 7 and 8. Throughput for both pairs decreases as the packet payload size increases. However, we clearly see that the faster the transmission rate the higher the throughput, despite the lower transmission probability we observed in Figs. 5 and 6. Quite clearly the fast transmission rates allow more data to be sent in less time, despite the apparent inefficiencies of the IFS at higher rates.

Finally we consider the probability of collision in Figs. 9 and 10. Again we see very similar profiles for 802.11b and 802.11g. Here we observe that the probability of collision is much greater for slow transmission speeds, which also helps to explain some of the lower throughput we observed in Figs. 7 and 8. Slightly counter-intuitively the collision probability reduces as frame length increases. One might think that longer transmissions are more likely to be interrupted by a transmission from a hidden node, but this does not seem to be the case. One reason for this is that at this high load more short frames are being transmitted than long ones, so there are more frames which can collide. This effect is particularly noticeable when the transmission speed is relatively low. When the transmission speed is high the ratio between transmission and waiting (IFS) is relatively low (hence the lower channel utilisation that we already observed), hence there is more time when the other node is not transmitting and so the chance of collision is reduced. In this case the difference between long and short frames makes much less difference than when the transmission rate is low.

6   Conclusion

In this paper, we have analysed the performance modelling in the IEEE 802.11b/g using PEPA by studying on the hidden nodes scenario. These results help us to better understand the performance of these protocols. In our scenario each node attempts to transmit whenever it is able, but collision occurs with a proportion of messages because the nodes are hidden from each other. Hence the maximum throughput is limited by the occurrence of collisions, the efficiency of the backoff process and the need to retransmit data and acknowledgements. The waiting times introduced by the IFS are tuned to work for the slowest transmission speeds in each version of the protocol. As such the maximum utilisation is achieved when the transmission is slowest. However, we also observe that slow transmission results in more collisions and hence the maximum throughput is far greater when the transmission rate is faster. In essence, faster transmission allows more data to be transmitted in less time with fewer collisions. Faster transmission is also shown to be less susceptible to variation in the collision probability with frame size. This observation leads us to speculate whether a lower collision rate might be achieved for slow transmission rates if the IFS periods were longer. This remains a question for future investigation.

In the next obvious future work, we will consider to study additional topological scenarios with more recent wireless protocols. The next step is to study IEEE 802.11n, including measures aimed at reducing the use of inter-frame spacing to increase its performance. We have observed here and in our previous work on 802.11b and g, inter-frame spacing is advantageous at reducing collisions and promoting fairness. Hence it seems reasonable to speculate that reducing the use of inter-frame spacing might in fact increase collisions from hidden nodes and increase problems of unfairness. However we may also speculate that (as we have observed here) overall throughput might be increased.

Acknowledgements. The financial support by Kurdistan Regional Government and Newcastle University is highly appreciated. The authors wish to thank Leïla Kloul for her generous help in the early stages of this work.

References

1. IEEE draft standard for IT-telecommunications and information exchange between systems local and metropolitan area networks-specific requirements part 11: wireless LAN medium access control (MAC) and physical layer (PHY) specifications, pp. 1–3701, February 2015
2. Abdullah, C.O., Thomas, N.: Formal performance modelling and analysis of IEEE 802.11 wireless LAN protocols. In: UK Performance Engineering Workshopp (2015)
3. Abdullah, C.O., Thomas, N.: Modelling unfairness in IEEE 802.11g networks with variable frame length. In: Wittevrongel, S., Phung-Duc, T. (eds.) ASMTA 2016. LNCS, vol. 9845, pp. 223–238. Springer, Heidelberg (2016). doi:10.1007/978-3-319-43904-4_16
4. Abdullah, C.O., Thomas, N.: Performance modelling of IEEE 802.11g wireless LAN protocols. In: Berlin 9th EAI International Conference on Performance Evaluation Methodologies and Tools (2015)
5. Alekhya, L., Mounika, B., Jyothi, E., Bhandari, B.N.: A waiting-time based backoff algorithm in the IEEE 802.11 based wireless networks. In: National Conference on Communications (NCC), pp. 1–5 (2012)
6. Anita, Singh, R., Priyanka, Indu: Performance analysis of IEEE 802.11 in the presence of hidden terminal for wireless networks. In: Jain, L.C., Behera, H.S., Mandal, J.K., Mohapatra, D.P. (eds.) Computational Intelligence in Data Mining-Volume 1, pp. 665–676. Springer, New Delhi (2015)
7. Argent-Katwala, A., Bradley, J.T., Geisweiller, N., Gilmore, S.T., Thomas, N.: Modelling tools and techniques for the performance analysis of wireless protocols. Adv. Wirel. Netw.: Perform. Model. Anal. Enhancement 3, 3–39 (2008)
8. Duda, A.: Understanding the performance of 802.11 networks. In: Proceedings of the 19th International Symposium on Personal, Indoor and Mobile Radio Communications, vol. 8, pp. 1–6 (2008)
9. Ekici, O., Yongacoglu, A.: IEEE 802.11a throughput performance with hidden nodes. IEEE Commun. Lett. **12**(6), 465–467 (2008)
10. Fiehe, S., Riihijärvi, J., Mähönen, P.: Experimental study on performance of IEEE 802.11n and impact of interferers on the 2.4 GHz ISM band. In: Proceedings of the 6th International Wireless Communications and Mobile Computing Conference, pp. 47–51 (2010)
11. Galloway, M.: Performance measurements of coexisting IEEE 802.11 g/n networks. In: Proceedings of the 49th Annual Southeast Regional Conference, pp. 173–178. ACM (2011)
12. Gilmore, S., Hillston, J.: The PEPA workbench: a tool to support a process algebra-based approach to performance modelling. In: Computer Performance Evaluation Modelling Techniques and Tools, pp. 353–368 (1994)
13. Hiertz, G.R., Denteneer, D., Stibor, L., Zang, Y., Costa, X.P., Walke, B.: The IEEE 802.11 universe. IEEE Commun. Mag. **48**(1), 62–70 (2010)
14. Hillston, J.: A Compositional Approach to Performance Modelling. Cambridge University Press, Cambridge (2008)
15. Hou, T., Tsao, L., Liu, H.: Analyzing the throughput of IEEE 802.11 DCF scheme with hidden nodes. In: 2003 IEEE 58th Vehicular Technology Conference, VTC-Fall, vol. 5, pp. 2870–2874 (2003)
16. Hou, T., Tsao, L., Liu, H.: Throughput analysis of the IEEE 802.11 DCF scheme in multi-hop ad hoc networks. In: Proceedings of the International Conference on Wireless Networks, pp. 653–659 (2003)

17. Khanduri, R., Rattan, S., Uniyal, A.: Understanding the features of IEEE 802. 11g in high data rate wireless LANs. Int. J. Comput. Appl. **64**(8), 1–5 (2013)
18. Kloul, L., Valois, F.: Investigating unfairness scenarios in MANET using 802.11b. In: Proceedings of the 2nd ACM International Workshop on Performance Evaluation of Wireless Ad Hoc, Sensor, and Ubiquitous Networks (2005)
19. Kumar, P., Krishnan, A.: Throughput analysis of the IEEE 802.11 distributed coordination function considering capture effects. Int. J. Autom. Comput. **8**, 236–243 (2011)
20. Kuptsov, D., Nechaev, B., Lukyanenko, A., Gurtov, A.: How penalty leads to improvement: a measurement study of wireless backoff in IEEE 802.11 networks. Comput. Netw. **75**, 37–57 (2014)
21. Medepalli, K., Gopalakrishnan, P., Famolari, D., Kodama, T.: Voice capacity of IEEE 802.11 b, 802.11 a and 802.11 g wireless LANs. In: Global Telecommunications Conference, GLOBECOM 2004, vol. 3, pp. 1549–1553. IEEE (2004)
22. Pham, D., Sekercioglu, Y.A., Egan, G.K.: Performance of IEEE 802.11b wireless links: an experimental study. In: Proceedings of the IEEE Region 10 Conference (TENCON) (2005)
23. Sandra, S., Miguel, G.P., Carlos, T.R., Jaime, L.: WLAN IEEE 802.11 a/b/g/n indoor coverage, interference performance study. Int. J. Adv. Netw. Serv. **4**, 209–222 (2011)
24. Sridhar, K.N., Ciobanu, G.: Describing IEEE 802.11 wireless mechanisms by using the π-calculus and performance evaluation process algebra. In: Núñez, M., Maamar, Z., Pelayo, F.L., Pousttchi, K., Rubio, F. (eds.) FORTE 2004. LNCS, vol. 3236, pp. 233–247. Springer, Heidelberg (2004)
25. Vucinic, M., Tourancheau, B., Duda, A.: Simulation of a backward compatible IEEE 802.11g network: access delay and throughput performance degradation. In: Mediterranean Conference on Embedded Computing (MECO), pp. 190–195 (2012)
26. Wang, Y., Garcia-Luna-Aceves, J.: Modeling of collision avoidance protocols in single-channel multihop wireless networks. Wirel. Netw. **10**(5), 495–506 (2004)
27. Xi, S., Kun, X., Jian, W., Jintong, L.: Performance analysis of medium access control protocol for IEEE 802.11g-over-fiber networks. China Commun. **10**(1), 81–92 (2013)
28. Xiao, Y.: IEEE 802.11n: enhancements for higher throughput in wireless LANs. IEEE Wirel. Commun. **12**(6), 82–91 (2005)
29. Younes, O., Thomas, N.: An SRN model of the IEEE 802.11 DCF MAC protocol in multi-hop ad hoc networks with hidden nodes. Comput. J. **54**(6), 875–893 (2011)

Simulation of Runtime Performance of Big Data Workflows on the Cloud

Faris Llwaah$^{(\boxtimes)}$, Jacek Cała, and Nigel Thomas

Newcastle University, Newcastle upon Tyne, UK
{f.llwaah,jacek.cala,nigel.thomas}@ncl.ac.uk

Abstract. Big data analysis has become a vital tool in many disciplines. Due to its intensive nature, big data analysis is often performed in cloud computing environments. Cloud computing offers the potential for large scale parallelism and scalable provision. However, determining an optimal deployment can be an expensive operation and therefore some form of prediction of performance prior to deployment would be extremely useful. In this paper we explore the deployment of one complex such problem, the NGS pipeline. We use provenance execution data to populate models simulated in WorkflowSim and CloudSim. This allows us to explore different scenarios for runtime properties.

Keywords: Big-Data · Scalability · NGS pipeline · WorkflowSim · CloudSim

1 Introduction

A big data workflow is composed of many applications that may involve large input data sets and produce large amounts of data as an output [5]. The scale and demand of these applications is such that they might rapidly overwhelm stand alone computing systems. One solution to this problem is to deploy the workflow into a commercial cloud environment, which provides ample resources and elastic provision. However, hiring resources clearly costs money and the process of tuning the deployment to ensure sufficient and efficient use of resources can be a costly exercise in itself. Therefore some means of predicting the performance of deployed workflows would be extremely useful and could save money.

In this paper, we explore this problem by considering a complex genomics data processing Next Generation Sequencing (NGS) workflow-based pipeline deployed on the Microsoft Azure public cloud [1,2]. The NGS pipeline is used to discover variants in patients exome. The local deployment of this pipeline, processing a cohort of 24 patient samples, typically takes several days to execute. The Azure deployment can potentially run much faster, but given limited funds it is necessary to find an optimal or near-optimal deployment which minimises both execution time and cost.

Fortunately a number of simulation tools have become available in recent years which enable a workflow to be simulated in repeatable and reproducible

© Springer International Publishing AG 2016
D. Fiems et al. (Eds.): EPEW 2016, LNCS 9951, pp. 141–155, 2016.
DOI: 10.1007/978-3-319-46433-6_10

experiments, with no charge for testing environment [9]. These simulators have been a significant tool for the evaluation and improvement a single workflow [3], although there is a lack of support for simulating a pipeline (a set of workflows).

In this paper, we have modified a simulation platform to simulate the execution behavior of the NGS pipeline as implemented in e-Science Central (e-SC). The main contributions of this paper is to propose a methodology for predicting the runtime and output data size using WorkflowSim/CloudSim, parameterised with realistic data from archived provenance file of e-Science Central workflows. In order to achieve this we have translated the e-SC workflow enactment model into a Pegasus workflow suitable for input into WorkflowSim and used WorkflowSim and CloudSim to predict runtime and the output data size. To the best of our knowledge this kind of prediction is novel.

The remainder of this paper is structured as follows: the next section covers some background and related work. Section 3 presents the simulation model. Section 4 discusses the proposed prediction methodology which is followed by Sect. 5 covering the evaluation. Finally, conclusions and future work are presented in Sect. 6.

2   Background and Related Work

Although providing cloud runtime estimation for big data workflows is a problem of significant interest, very few studies are currently available in the literature. This is partially, due to the complexity of the problem in terms of workflow performance behavior and due to the modernity of cloud simulation. Some notable contributions in this area include Rak *et al.* [11], who presents a technique to evaluate the trade-off between costs and performance of the cloud application through benchmarks and simulation based on the mOSAIC framework. In [13] the authors extend this approach to consider *bag-of-tasks* scientific applications. The integrated framework with the cloud simulation environment is able to predict the behavior of development stage performance and cloud resource usage. Rozinat *et al.* [12], describe a simulation system for operational decisions to support the context of workflow management. The proposed approach combines and extends the workflow management system YAWL and the process mining framework ProM. CloudProphet [8] aims to predict the performance and costs of legacy applications when executed on cloud infrastructures. The advantage of this approach is focused on applications cloud-aware by design, which means it takes into account the elasticity of elasticity rather than using a framework to predict the performance. The framework presented in [14] for performance prediction of parallel programs on hierarchical clusters is based on two principle steps:- one at the installation time of the parallel application and the other at the runtime. In order to model accurately the components, they are sketched those components. In the second step in this approach the generated model was used to the completion time estimation via the fast simulator MPI-PERF-SIM.

Our approach is built on to of WorkflowSim [4], which is an extension of the CloudSim simulator. This is done by providing multiple layers on top of the

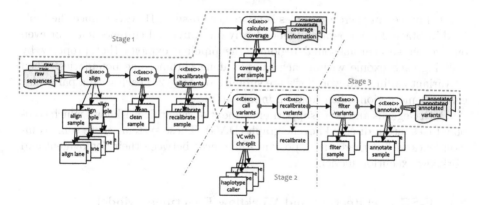

Fig. 1. The structure of the NGS pipeline; highlighted in dashed blue is the top-level workflow; red dots indicate the synchronisation points when the top-level invocation waits for all child invocations to complete. (Color figure online)

existing task scheduling layer of CloudSim, such as workflow mapper, workflow engine, clustering engine, and workflow scheduler. Cloudsim [3] provides the realistic components such as data centre, host, policies and workloads.

3   NGS Pipeline Simulation

To explore the problem of simulating workflow deployment in the cloud, we have chosen a case study using the NGS pipeline [1]. These workflows were designed following the WES data processing pipelines used at the Institute for Genetic Medicine, Newcastle University. In general, a pipeline consists of a composition of workflows that include typical NGS processing steps [10], which are raw sequence alignment (BWA), cleaning (Picard), sequence recalibration, filtering, variant calling and recalibration (GATK), coverage analysis (bedTools), and annotation (Annovar). It consists of a top level, coordinating workflow that invokes 8 sub-workflows, each of which implements one step of the pipeline, see Fig. 1.

For each step, the sub-workflows are executed synchronously in parallel over a number of samples or sub-chromosomal regions. Each patient sample includes 2-lane, pair-end raw sequence reads. The average size of input is about 150 Gbases per sample, which is provided as compressed files of nearly 15 GiB size. The top-level workflow processes N of these samples ($N \geq 6$) by submitting N sub-workflow invocations for a particular step. Then, it waits until all of them complete, and moves on to the following step.

Overall, the pipeline involves three key stages: (1) preparation of the raw sequences for variant discovery and coverage calculation, (2) variant calling and recalibration, (3) variant filtering and annotation. Stages 1 and 3 are executed in a loop so that all tools involved are invoked on each sample separately. As there is no dependency between samples in these two stages, paralellisation is straightforward. Conversely, Stage 2 runs only once for all input samples, thus parallel processing across samples is no longer possible. However, since the tools used in Stage 2 can operate independently on individual chromosomes (or even on smaller sub-chromosomal regions), the pipeline exploits that property by splitting each exome within each sample along chromosome boundaries. Then M variant calling sub-workflows are submitted ($M \geq 23$) each with a particular chromosome region taken from all input samples.

Both Pegasus and e-SC support enactment of scientific workflows which combine tasks into a directed acyclic graph (DAG). These systems share some common features but there are important differences between their deployment and workflow execution model.

3.1 E-SC Architecture and Workflow Enactment Model

e-SC consists of three main components: the server, database and workflow engine. It follows the common *master-worker* pattern in which the server orchestrates execution of workflows across one or more workflow engines. All e-SC components can be deployed on a single VM (*all-in-one deployment*) but in larger scale experiments, such as the NGS pipeline, they are deployed separately with single server and database VM and multiple engine VMs. The e-SC workflow enactment model is based on the *work stealing* approach: the server submits workflow invocations to a shared FIFO queue. From there invocations are pulled by the engines. Each engine can run one or more invocations concurrently in order to improve performance on a multi-core VMs.

The e-SC workflows can be of two types: basic and compound. Basic workflows execute within a single engine (within a single invocation thread on that engine), and so the data transfer between tasks is enclosed within a VM and can be very efficient. In addition there are compound workflows, which are workflows which submit one or more subworkflows. A subworkflow can again be compound or basic. Of course, data transfer between the parent and its child subworkflows is supported by the server. However, in the Cloud, workflow engines can directly communicate with scalable cloud storage such as Azure Blob Store or Amazon S3, which enables effective data transfer for large scale workflow applications. Moreover, links between blocks can transmit a list of input data and so a single parent workflow can start multiple subworkflows – one for each element on the list. Figure 1 shows the architecture of e-SC deployed in the Azure Cloud.

3.2 Pegasus Architecture and Workflow Enactment Model

WorkflowSim follows the execution model of the Pegasus WfMS. In Pegasus a workflow consists of tasks, each of which represents a node, and the task dependencies, denoted by the edges. Often, a workflow is modeled as a DAX to DAX relationship; we assume that $DAG = (V, A)$, where set of vectors $V = \{T_1, T_2, \ldots, T_n\}$ represents tasks in the workflow and set of arcs A represents data dependencies between these tasks. Moreover, data transfer between tasks is achieved using Condor File IO in the case of a non-shared file system setup.

3.3 Modelling the Pipeline in WorkflowSim

Because, e-SC workflows can represent combinations of more fine-grained tasks and also due to the difference between the workflow model and possible invocation trace, we had to find a way to map an e-SC workflow into one that WorkflowSim could enact. The chosen approach was to represent the actual invocation trace of an e-SC workflow as a compatible Pegasus workflow, which could be done for the NGS pipeline using the provenance logs provided by e-SC. The provenance logs allow us to trace the complete graph of tasks and workflows that were involved in producing a specific output. They also include the block execution time and the amount of data transferred by each block. We used this data to reconstruct our NGS pipeline workflow as a WorkflowSim workflow. Each task in Pegasus may run on different VM (unless there's clustering turned on) so we decided to model e-SC workflows as WorkflowSim tasks. This seems to be an appropriate abstraction level, however subworkflows in e-SC can be enacted in the middle of the parent workflow, so we had to split every e-SC workflow into parts connected by the subworkflow submission blocks; this is depicted in Fig. 1. Following this approach we were able to map an execution trace of the NGS pipeline ran on e-SC as a WorkflowSim workflow descriptor. Figure 2 illustrates the mapping. Note that for a different number of patient samples in the batch there is a differently sized DAX (DAG in XML) descriptor.

4 The Prediction Methodology

WorkflowSim requires as input a description of the execution environments, the workflow to be executed and the execution times for each task. WorkflowSim then simulates the deployment of the workflow on the environment. We wish to use this simulation to predict the performance of the NGS pipeline with different sample input sizes. From the above we can provide WorkflowSim with an input model of the NGS pipeline. However, as yet we have no idea about the task execution times or the parameters of the execution environment (MIPs and bandwidth). Our approach is therefore to gather provenance data from sample executions with small number of input samples, use this data as input to WorkflowSim to estimate the execution environment parameters and then use this to gain predictions for larger sample sizes from further WorkflowSim simulations. These predictions can then be used to find the best selection of resources (e.g. number of VMs) on which to deploy the pipeline for larger sample sizes.

4.1 Preparation Phase

(i) Providing Parameters to WorkflowSim: A WorkflowSim toolkit must capture a complete description of the tasks, such as identification, runtime, input data sizes, and output data size. One of the trends is to make a WorkflowSim automation toolkit by providing it with an opportunity to predict the transformation parameters such as runtime and output data size. Therefore, this facility will help a user to use WorkflowSim in an easy and efficient way. As such, the NGS pipeline is launched only with the input data sizes. This will require using two prediction models and integrating them into WorkflowSim to perform this job.

(ii) **WorkflowSim Input Compatibility:** This step of the preparation phase is related with the conversion from e-SC provenance traces to WorkflowSim workflow model, in order to obtain the schema of NGS pipeline that will be accepted as input workflow to the WorkflowSim.

Based on the workflow execution model description above, we convert the NGS pipeline from an execution model in e-SC Central to an execution model in Pegasus. Figure 2 shows the result of a hierarchical pipeline graph which can be represented by DAG implementing a single NGS pipeline that represents one possible execution path of the workflow schema, where an execution consists of all nodes and edges within a workflow DAX beginning from the start node (i.e. 1-sample input) to the end node. For running 6 samples NGS pipeline, the implementation will be within 6 paths of the workflow schema (i.e. 6-sample input), and so on for running N NGS will be implemented within N paths.

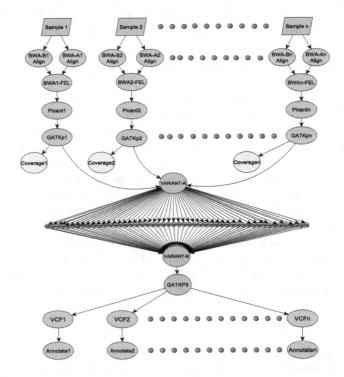

Fig. 2. Invocation graph of the NGS pipeline with N samples.

(iii) **Data Submission Format:** Here we describe the workflow and present a new XML file format for workflow states that enable the WorkflowSim to interface NGS pipeline in an acceptable way. As we mentioned above, the original parser model of the WorkflowSim has been modified to analyse and parse a new XML file of the pipeline. In order to execute the e-SC workflow application in our modified WorkflowSim, workflows are described by users manually as DAXs, where the node represents individual workflow/subworkflow, and the edges represent execution dependencies between the (sub)workflows. Figure 2 illustrates the abstract description abstract as an XML file to represent the whole pipeline, which captures all logical identifiers with which the task should be invoked, such as input/output file, task identifier (id), and required run time (runtime). For example, the following is a brief record on XML file that represents the BWA_A1 Align workflow.

```
id = BWA_A1_AL;
file = InputDir_BWA_A1_AL.dat;  size="7507423824";
file = BWA__A1_AL.out;          size = <->;
runtime = <->
```

In the case of size parameters, each is assigned by value of input sample, the second size and runtime parameters should be assigned by the prediction models.

4.2 Building the Prediction Models Phase

In this phase, we describe the method to build a prediction model. To achieve this, we need the following steps:

(i) **Data Collection and Feature Set:** The main task for this step is to shed light on the predicting modules by collecting information from the provenance file for each invocation which is needed for performance prediction. Thus, we use historical data, including the details information about the execution of the workflows in the pipeline such as invocation Id, Workflow Name, Block number, Block name, Start time, End time, Input data size, and Output data size. The following steps describe a method of extracting the above parameters from analytical provenance file:

- Sorting the information by an invocation Id and Start time. To obtain an ordering of the pipeline blocks as were executed in a real cloud. This facility helps us easily extract the actual execution time of each block.
- Extracting the runtime for each block, (i.e. runtime = End time − Start time).
- Specifying the input and output data volume of each block.

Table 1 gives an example of the provenance data pertaining to the runtime and (input-output) data volume which is collected from an execution before the actual simulation of a workflow is started. All extracted parameters play a significant role in our experiments for time execution estimation of the task. Moreover, to derive the output data volume that will be used as input for next task.

Table 1. Basic characteristics of a selection of tasks of a 10-sample pipeline execution extracted from its provenance trace.

Pipeline step	Input data [MB]	Output data [MB]	Run time [s]
BWA1_FEL	15,862	11,336	18,807
BWA_A1_AL	7,507	7,963	9,871
PICARD1	11,342	7,411	8,941
GATKP1_1	7,417	2,766	28,703
VARIANTA	344	344	23,792
GATK phase3	55	43	943
VCF1	55	43	175
COVERAGE1	2,766	16	280
ANNOTATE1	43	204	1,206

(ii) Extracting Prediction Equations: To run NGS pipeline tasks, as in Fig. 2, on WorkflowSim we need to know the execution time of each task before the submission is done. This aim prompted us to build prediction models for both runtime and data output sizes which are based on live performance data and using a statistical prediction approach for extracting the equations of prediction. The time parameters can be extracted from input data volume for each block and output data volume parameters. For the majority of the tasks that shown in Fig. 2, it is possible to generate the prediction equations to estimate execution time of each task based upon the size of the input data. For example, Fig. 3 shows a prediction equation of the BWA1_FEL for three 6-sample input using a simple linear regression model. The same method was used to obtain further equations of the output data sizes.

However, using data input size to generate estimation equations was inadequate for the Haplotype workflows which are the entry-point to the part of the pipeline that runs under the chromosome-split regime. It means that each Haplotype workflow as the input uses data of all patient samples but is configured to read different chromosomal region of them. Thus, we used the region length as a division factor for different input sizes.

(iii) Set Up Training Sets: From measurements we have obtained three data sets based on different numbers of input samples (6, 10 and 12 samples executes 3, 2 and 2 times each respectively). Each set can be used to train an input sample for execution environment parameter estimation. Thus, we can create different sized training sets for time prediction of a scalable input sample. For example, if we have 12-samples input, the prediction model can use a training set based on 6 and/or 10 samples. In the evaluation we can explore whether having more data points across all available training sets (in this case 6 and 10) provides a better prediction than using just the data from one set (6 or 10). Ideally we would get

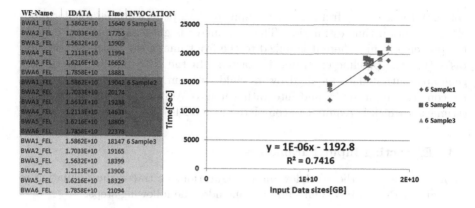

WF-Name	IDATA	Time	INVOCATION
BWA1_FEL	1.5862E+10	15640	6 Sample1
BWA2_FEL	1.7033E+10	17755	
BWA3_FEL	1.5632E+10	15909	
BWA4_FEL	1.2113E+10	11994	
BWA5_FEL	1.6216E+10	16652	
BWA6_FEL	1.7858E+10	18881	
BWA1_FEL	1.5862E+10	19042	6 Sample2
BWA2_FEL	1.7033E+10	20174	
BWA3_FEL	1.5632E+10	19238	
BWA4_FEL	1.2113E+10	14631	
BWA5_FEL	1.6216E+10	18805	
BWA6_FEL	1.7858E+10	22378	
BWA1_FEL	1.5862E+10	18147	6 Sample3
BWA2_FEL	1.7033E+10	19165	
BWA3_FEL	1.5632E+10	18399	
BWA4_FEL	1.2113E+10	13906	
BWA5_FEL	1.6216E+10	18329	
BWA6_FEL	1.7858E+10	21094	

Fig. 3. Linear model of equation prediction.

a good prediction from just using the 6 sample training set, as this would clearly be the cheapest to produce.

4.3 Integrating the Derived Equations Phase

In this phase, we integrate the prediction equations that have been built in the building phase into WorkflowSim by considering two issues. Firstly, a run time prediction should be given for each task in case of submitting the tasks with predefined execution time to the WorkflowSim. Secondly, the output data volume should be calculated, which is a passing a factor to other tasks during running the pipeline. So, we have constructed two estimation models:

(i) Runtime Prediction Model: One key benefit potential statistical prediction method has to match for parameters of input data to predict the parameters of the output data of task/job, making this prediction by using past information [7]. We have used the linear regression method as a solution to address the estimation. This approach manages the relationship between two variables, X is input variable (i.e. input data volume) and Y is dependent output variable (i.e. runtime), to extract Y from X.

Execution time prediction is an important factor in cloud computing and in simulation [6]. However, in a WorkflowSim, a task is assigned according to its size which is defined by the user. Therefore, it was necessary to develop a model in WorkflowSim for runtime estimation which is required to simulate the task.

(ii) Output Data Volume Prediction Model: The approach taken by runtime prediction method to generate a model is followed in this method as well. However, the difference lies in two considered variables, i.e. X is input variable (i.e. input data volume) and Y is dependent output variable (i.e. output data volume), to extract Y from X. The volume of data plays a crucial role for modelling execution time estimation. This parameter is gathered from real data in the provenance file where it is linked to the front line of the tasks (i.e. the first tasks that the pipeline execution is started). In the NGS pipeline, every task generates output data required by its child as input. This method is required for constructing a model and integrating it into WorkflowSim for output data estimation which is required to complete our models.

4.4   Extracting Input Parameters

In order to retrieve the WorkflowSim configuration, we traced the simulation to generate MIPs and BW. Our approach includes the following steps:

1. Selecting minimum and maximum value of MIPs and BW parameters.
2. Running the WorkflowSim individually for each sample depending on the chosen MIPs and BW values with defined range in step 1 to generate estimated runtime of pipeline execution.
3. Applying an error function to find the error value between real time and estimated time for each running input sample by implementing the following formula.

$$error ratio = \sum_{j=1}^{N} (RT - ET)^2. \tag{1}$$

Where RT and ET are Real time and Predicted time respectively. N is the number of input sample in one training set.
4. Repeat step 2 and 3 with fixed skip of MIPs and BW values across a given range and detect the minimum *error ratio*.

The input parameters have been generated from three scenarios as in Figs. 4, 5 and 6. Each scenario denotes one training set, i.e. scenario 1 denotes training set {6}, scenario 2 denotes training set {6 + 10}, and scenario 3 denotes training set {6 + 10 + 12}. Therefore, the above steps have been implemented on three scenarios to specifying input parameters value of the training sets.

5   Evaluation

This section presents our evaluation of NGS pipeline execution and its use with an adopted WorkflowSim to derive a scalability and performance optimization based on estimated run time. Firstly, we describe the experiment setup, then we present our results on the accuracy of the NGS pipeline running with three input samples, finally we discuss the evaluation of the estimated results for relative errors between different running samples to derive the expectation of a Big-Data samples.

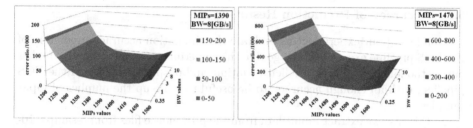

Fig. 4. Scenario 1 extracting parameters of 6-samples.

Fig. 5. Scenario 2 extracting parameters of $(6+10)$-samples.

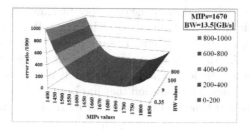

Fig. 6. Scenario 3 extracting parameters of $(6+10+12)$-samples.

5.1 Experiment Setup

The NGS pipeline is composed of 8 tasks for each path and between them there are 53 common workflows (VARIANT-A, HAPLOTYPE-CLEAR, VARIANT-B, and GATK-phase3). We ran the application with the size of Total Tasks = $N \times 9 + 53$, where N is the number of input samples. For example, if $N = 6$, then Total Tasks = $6 \times 8 + 53$. So, when we have 6 input samples, the experiment consists of 101 tasks. For each task we generate the estimation of the output data size and runtime using the simulation and then use the output data size to generate the input data size for the subsequent task. We configured WorkflowSim to simulate one datacentre and 12 virtual machines (VMs) to represent four VMs in the real cloud, each with three execution threads. We allocated the capacity of the computation unit with MIPs and bandwidth (BW) values derived from simulating the training set. Moreover, for data transfer delay, a shared file system has been used for one datacentre, where, the data transfer time is already considered in the task execution time and there is a varying setting of a BW value depending on which training set we will use. The space shared mode of the VMs has been defined as only one VM can run one task at a time. We have used three different sizes of the input sets with 6, 10, and 12 patient samples, based on the data we have available for training and validation.

The size of the input sets was a trade-off between what is used in clinical practice (30–40 patient samples) and the cost it takes to run the pipeline in the Cloud. And the 6-sample input set was the minimal size for which the pipeline completed successfully.

5.2   Accuracy of Prediction Results

We conducted our experiment to evaluate the simulation accuracy prediction of the NGS pipeline execution. The goal of this experiment is to determine whether our methodology is able to predict a runtime and output data sizes of the NGS pipeline, with an enough accuracy to be useful to scale up the number of input samples.

In this case we extract input data from the provenence files from 6-sample 10-sample, and 12-sample executions and then use this to execute WorkflowSim over the same input scale to give an estimated execution time to compare with the real execution time from the provenance data. In this way, the implementing of pipeline to generate estimated run time on a specific training set that mentioned before. i.e. 6-sample would be trained on training set {6}, 10-sample would be trained on training set {6 + 10}, and 12-sample would be trained on training set {6 + 10 + 12}. As the real execution potentially vary with each run and WorkflowSim gives an single prediction, this predicted value will therefore be different to the real times. The equation that has been used to calculate a relative error between estimated time and real time for each case as follows:

$$RelativeError = \frac{\sum_{j=1}^{N} |RT - ET|}{\sum_{j=1}^{N} (RT)} \tag{2}$$

where RT and ET are Real time and Predicted time respectively. N is the number of input sample in one training set.

In each case the values of MIPs and BW which gives the minimum error over the training set are used to give the estimated time. The results in each case are shown in Fig. 7. As expected the errors are relatively small (<10%). It might be slightly counter-intuitive that the error for the 6-sample (0.090) is larger than that for either the 10-sample or 12-sample (0.071 and 0.075 respectively). However, this is due almost entirely to one outlier in the measured execution times which has a disproportionate effect on the 6-sample as there are fewer data elements in the training set. Such experimental variance could clearly be reduced by ignoring the outlying value. However, we have chosen not to manipulate our results in this way as (a) the outlying result is a genuine data point, and (b) our number of data points in each training set is so small that we have no statistical basis on which to say which result is an outlier and which is not. However, as an aside, we have executed the experiments without this data point and the results do improve considerably.

5.3   Relative Errors at Input Scalable

In the previous section, we verified the accuracy of the prediction model by comparing the actual and estimated runtime derived from the training sets. This process was self-reflective in that we included the provenance data from the sample size we were predicting within the training set. We now wish to consider a pure prediction of the 12-sample input by using the 6- and 10-sample

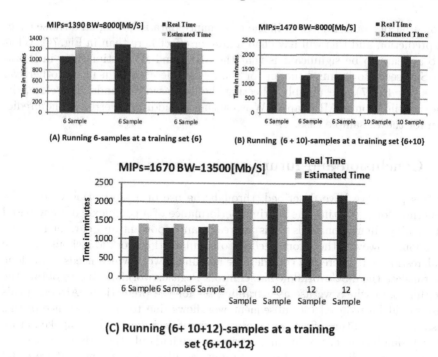

Fig. 7. The real and estimated time for different sizes of the training set.

sets as training data. This allows us to consider whether our approach can indeed give rise to a useful prediction in this scenario which might be used to procure infrastructure in the cloud. The results are shown in Fig. 8.

Using the 6-sample data as the training set for the 12-sample case gave a relative error of approximately 0.187 (18.7 %, see Fig. 8A). Using the 6- and 10-sample data as the training set for the 12-sample case gave a relative error of approximately 0.112 (11.2 %, see Fig. 8B). In both cases the predictions underestimate the execution time. In the case of the 6-sample training set we already

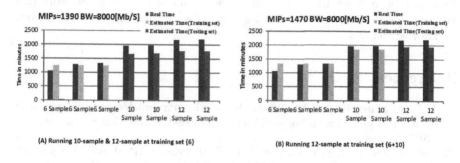

Fig. 8. The real and estimated time for different training and testing set sizes.

know that one outlying result in the provenance data is having an adverse effect on prediction and this will have had a greater effect here than in Fig. 7C. What also appears to be significant is that the MIPs and BW estimations in Work-flowSim seem to be much larger for the 12-sample case than in the other two cases (see Fig. 7), presumably reflecting the increased demands. Therefore, the under-estimation of MIPs and BW is causing an additional error in the prediction for the 12-sample case.

6   Conclusion and future work

In this paper we have described, through the use of a motivating case study, a method for predicting the runtime performance of a complex workflow based on existing simulation tools. This addresses an important and current research question. However, the effort needed to simulate this workflow should not be underestimated. There is a considerable amount of work which needs to be done to translate the e-SC workflow model into a Pegasus workflow and to obtain the baseline data which we use as a training set for the predictions. Although this cost would be reduced for subsequent workflows due to the experience of this case study, the future effort would also be significant due to the bespoke nature of each simulation set up. In our experiments we had only three data sets to consider, corresponding to 6, 10 and 12 input samples. This meant that we could validate predictions for 10 samples, using the 6 sample data as training, and for 12 samples, using the 6 and/or 10 samples for training. Ideally, the 6 sample data would have proved sufficient for good prediction as this would enable a relatively cheap means of data collection. However, the results clearly show that for 12 samples the addition of the 10 sample data offers better training. It is disappointing that the results are not more convincing. While a reasonable level of accuracy has been shown to be achieved, a similar level of accuracy could potentially be obtained by deriving execution data at a number of scales and performing a linear extrapolation. However, one advantage of our simualation approach is that we are able to make predictions based on a single set of observations, which is clearly less costly in terms of access to the cloud provider.

The results in this paper relate to one complex workflow from one application domain. Clearly if we are to draw any general conclusions from this work then we need to conduct more case studies with different applications with different workflow structures.

References

1. Cała, J., Marei, E., Xu, Y., Takeda, K., Missier, P.: Scalable and efficient whole-exome data processing using workflows on the cloud. Future Gener. Comput. Syst. (2016, in press)
2. Cała, J., Xu, Y., Wijaya, E., Missier, P.: From scripted HPC-based NGS pipelines to workflows on the cloud. In: 2014 14th IEEE/ACM International Symposium on Cluster, Cloud and Grid (2014)

3. Calheiros, R., Ranjan, R., Beloglazov, A., De Rose, C., Buyya, R.: CloudSim: a toolkit for modeling and simulation of cloud computing environments and evaluation of resource provisioning algorithms. Softw.: Pract. Exp. **41**, 23–50 (2010)
4. Chen, W., Deelman, E.: WorkflowSim: a toolkit for simulating scientific workflows in distributed environments. In: 2012 IEEE 8th International Conference on E-Science (2012)
5. Deelman, E., Gil, Y.: Managing large-scale scientific workflows in distributed environments: experiences and challenges. In: 2006 Second IEEE International Conference on e-Science and Grid Computing (e-Science 2006) (2006)
6. Fan, C., Chang, Y., Wang, W., Yuan, S.: Execution time prediction using rough set theory in hybrid cloud. In: 2012 9th International Conference on Ubiquitous Intelligence and Computing and 9th International Conference on Autonomic and Trusted Computing (2012)
7. Iverson, M., Ozguner, F., Potter, L.: Statistical prediction of task execution times through analytic benchmarking for scheduling in a heterogeneous environment. IEEE Trans. Comput. **48**, 1374–1379 (1999)
8. Li, A., Zong, X., Kandula, S., Yang, X., Zhang, M.: CloudProphet. ACM SIG-COMM Comput. Commun. Rev. **41**, 426 (2011)
9. Long, W., Yuqing, L., Qingxin, X.: Using CloudSim to model and simulate cloud computing environment. In: 2013 Ninth International Conference on Computational Intelligence and Security (2013)
10. Pabinger, S., Dander, A., Fischer, M., Snajder, R., Sperk, M., Efremova, M., Krabichler, B., Speicher, M., Zschocke, J., Trajanoski, Z.: A survey of tools for variant analysis of next-generation genome sequencing data. Brief. Bioinform. **15**, 256–278 (2014)
11. Rak, M., Cuomo, A., Villano, U.: Cost/performance evaluation for cloud applications using simulation. In: 2013 Workshops on Enabling Technologies: Infrastructure for Collaborative Enterprises (2013)
12. Rozinat, A., Wynn, M.T., van der Aalst, W.M.P., ter Hofstede, A.H.M., Fidge, C.J.: Workflow simulation for operational decision support using design, historic and state information. In: Dumas, M., Reichert, M., Shan, M.-C. (eds.) BPM 2008. LNCS, vol. 5240, pp. 196–211. Springer, Heidelberg (2008)
13. Rak, M., Turtur, M., Villano, U.: Early prediction of the cost of HPC application execution in the cloud. In: 2014 16th International Symposium on Symbolic and Numeric Algorithms for Scientific Computing (2015)
14. Achour, S., Ammar, M., Khmili, B., Nasri, W.: MPI-PERF-SIM: towards an automatic performance prediction tool of MPI programs on hierarchical clusters. In: 2011 19th International Euromicro Conference on Parallel, Distributed and Network-Based Processing (2011)

Modeling and Analysis of Human Behavior

Modeling Human Decisions in Performance and Dependability Models

Peter Buchholz[1], Iryna Felko[1], Jan Kriege[1(✉)], and Gerhard Rinkenauer[2]

[1] Informatik IV, TU Dortmund, 44221 Dortmund, Germany
{peter.buchholz,iryna.felko,jan.kriege}@cs.tu-dortmund.de
[2] Leibniz Research Centre for Working Environment and Human Factors,
44139 Dortmund, Germany
rinkenauer@ifado.de

Abstract. Many systems are driven partially by human operators who decide about basic operations that influence system behavior. Therefore the performance and dependability depend on the technical system and the human operator. Performance and dependability models usually include a detailed model of the technical infrastructure but the human decision maker is only roughly modeled by simple probabilities or delays. However, in psychology much more sophisticated models of human decision making exist. For tasks with two choices usually diffusion models are applied. These models include information about the process of human decision making based on perception or memory retrieval and take into account the time pressure under which decisions have to be made. In this paper we combine these diffusion models with Markov models for performance and dependability analysis. By using a discretization approach for the diffusion model the combined model is a Markov chain which can be analyzed with standard means. The approach allows one to integrate detailed models of human two-way decisions in performance and dependability models.

Keywords: Markov models · Human decision making · Numerical analysis

1 Introduction

In many technical systems decisions are at least partially made by humans who supervise or control the operations. Examples of those systems in computer science are data centers, supercomputers or large computer infrastructures where humans control decisions about resource allocation, maintenance or capacity assignment. The systems are a specific type of a socio-technical system.

For the analysis of this kind of system, both aspects, the function of the technical part as well as the behavior of the human decision maker have to be taken into account. In both cases, stochastic models seem to be adequate to describe the behavior. Nevertheless, there are significant differences in modeling

© Springer International Publishing AG 2016
D. Fiems et al. (Eds.): EPEW 2016, LNCS 9951, pp. 159–173, 2016.
DOI: 10.1007/978-3-319-46433-6_11

the different aspects. Performance and dependability models are based on discrete events and often Markov or semi Markov models [22] are applied to allow a numerical or even better analytical evaluation of system behavior. Human decision making is often based on perception which means that the human decision maker collects information about the state of her or his environment and this information drives the decision in a specific direction with some probability. The basic model for decisions with two choices are diffusion models [15] which are stochastic fluid models. These models can only be analyzed via simulation. However, by discretizing the continuous state space of the fluid model, it can be approximated by a discrete time [6,25] or even continuous time [3] absorbing Markov chain of the birth death type. In this way both model types, performance/dependability models and human decision models, become similar and, in principle, can be integrated.

The integration of human decisions in performance/dependability models is not really advanced. Only in agent based simulation models both parts are integrated [26]. However, in agent based models usually very detailed human behavior models and performance models are integrated which are only amenable to simulative analysis and which often do not support the understanding of the processes in the real system. In more abstract models, which are often analyzed numerically, human behavior models and performance/dependability models have, to the best of our knowledge, not been integrated yet.

In this paper we propose an approach to combine the diffusion model for two-choice decision making of humans and standard performance/dependability models. It is shown that both models are coupled, resulting in a Markov model which can be analyzed numerically, if the state space size permits, or by simulation. Even more, human decisions can be based on the state of the performance/dependability model resulting in multi-stage decision making in the sense of the approach presented in [7]. This paper focuses on the model perspective, it does not consider in detail the fitting of model parameters or the solution of the resulting models. For both steps methods are available. However, in particular for parameter fitting, research work is necessary to assure that performance/dependability model and human decision model are of similar accuracy.

The paper is structured as follows. In the following section we review related work. Section 3 introduces models of human decision making, Afterwards, in Sect. 4, the combination of performance/dependability and human decision making is introduced. Then we present some example models. The paper ends with the conclusions where we also give an outlook of future research.

2   Related Work

Quantitative analysis of systems like computer and communication networks, manufacturing plants or logistics networks is often done using discrete event models that, depending on the level of detail, can be analyzed by simulation or numerically [22]. Many systems involve at least some basic level of human interaction [17] due to decisions made by human operators or failures caused by human errors. For a realistic modeling of the system this human interaction has to be taken into account as well and several techniques have been proposed in the past to include human decisions into performance and dependability models.

The authors in [29] identified four approaches to model human decision makers in models of manufacturing systems. In the simplest case the decision maker is only modeled as a passive resource, implying that the decision operation requires a certain amount of time, but the conclusion triggers no reaction in the model, i.e. the outcome is given by predefined probabilities. More elaborate techniques use either global or local rules or depend on cooperation relationships as in agent-based models.

Many older approaches try to combine discrete-event simulation with ideas from artificial intelligence to model human-to-system interaction. Various approaches linked a simulation model with rule-based expert systems that are used to represent the decision maker for different application scenarios like simulation of a coal yard [8], replenishment at sea in the Royal Navy [27], harbor unloading by a steel company [12] or allocating lorries at a loading bay [18]. Another example from artificial intelligence are neural networks that were used in [4] for representing human decision making. The approaches from artificial intelligence can also be part of the knowledge-based improvement methodology as described in [17] where visual interactive simulation is used to collect example decisions from a human decision maker which are used to parametrize for example an expert system.

More recently the focus shifted to agent-based approaches when modeling human behavior. For example a model of human operators using the Belief-Desire-Intention agent paradigm to handle unexpected situations in an automated shop floor control system is proposed in [28]. The authors in [10,11] also used the Belief-Desire-Intention paradigm to model evacuation behavior of humans and to evaluate different evacuation performance indicators. In most papers like e.g. [13] the implementation of the agent behavior is explained in detail, but the agents are not really part of a larger model of a technical system that is influenced by the agents' decisions or only simple scenarios for the agents are sketched [19].

The approaches mentioned so far have in common that they require a simulative analysis to obtain performance results. In Markov models, that can be analyzed numerically or analytically, human behavior has only been modeled at a very abstract level. For example [5,20] present Markov models that contain common (i.e. technical) failures and human errors that lead to a nonfunctional system. However, the human errors are in fact only modeled by a simple transition with the average rate of human errors.

To the best of our knowledge diffusion models [15] that are a common concept to model decisions with two choices have not been used as part of a performance model with human decision makers, although, as already mentioned, they have the big advantage that they can be used as part of Markov models by discretization of the state space. Diffusion models are widely used in cognitive sciences in application areas ranging from cognitive tasks, reinforcement learning and the field of neuro-physiological measures to analysis methods in clinical research to examine distinct diseases [16]. Diffusion models provide evidence according to the cognitive mechanism of the human decision-making deliberation process, e.g., the relationship between reaction times and probabilities of making the particular decision and external factors.

3   Models for Human Decisions

The basic model which is used for two way decisions is the diffusion model [15]. Let $x(t)$ be the state of the decision maker at time t $(t \geq 0)$ and $x(0)$ the known initial state. We denote the two possible decisions as -1 and 1. Evaluation of $x(t)$ is described by the following stochastic differential equation.

$$\frac{dx}{dt} = \mu(x(t), t) + \sigma(x(t), t) dW(t) \tag{1}$$

where $W(t)$ is a Wiener process, i.e. a stochastic process defined for $t \geq 0$ with $W(0) = 0$, such that the increments $W(t) - W(s), t > s$ are Gaussian with zero mean and variance $t - s$ (i.e. $W(t) - W(s) \sim N(0, t - s)$) and increments for nonoverlapping time intervals are independent. Thus,

$$x(t + \Delta) \approx x(t) + \mu(x(t), t)\Delta + \sigma(x(t), t)W(\Delta) \tag{2}$$

for some small Δ. The decision process is driven by the drift rate $\mu(.)$ which shows the general direction and a random term which is given by the outcome of the Wiener process multiplied with $\sigma(.)$. In the simplest version we can assume $\mu(x(t), t) = \mu$ and $\sigma(x(t), t) = \sigma$.

To model decision making, we define two thresholds $\theta_{-1} < 0$ and $\theta_1 > 0$ and assume that $x(0)$ is selected from the open interval (θ_{-1}, θ_1). Then the process evolves until one of the two thresholds is reached at some point in time t and is stopped, i.e. decision -1 is made at time t if $x(t) = \theta_{-1}$ and decision 1 is made at time t if $x(t) = \theta_1$. By the selection of θ_{-1} and θ_1 different situations can be modeled. It is assumed that the decision maker is able to control decision speed by varying the interval (θ_{-1}, θ_1), viz. if there is more time pressure (i.e. by instruction) the interval may be reduced to make faster decisions. However, by reducing the interval the influence of the random part due to the Wiener process becomes larger and this usually implies that the price for faster decisions are more false decisions. After a decision is made, the state is reset to $x(0)$ and $x(0)$ is selected according to some probability distribution π on (θ_{-1}, θ_1). If the initial state depends on the previous decision, we can define two probability distributions π_{-1} and π_1.

The process behavior according to (1) with constant parameters μ and σ can be simulated in different ways [25] and results may be used to estimate the probability of decisions -1 or 1 and of the distribution of the time to make decision -1 or 1, respectively.

To make the model more flexible than it is with constant parameters μ and σ, a multi-stage model has been introduced [7], then the general Eq. (1) becomes

$$\frac{dx}{dt} = \mu_k + \sigma_k dW(t) \tag{3}$$

where $k = 1, \ldots, K$ is the stage the decision maker is in. Later we will define the stage using the state of the Markov model that describes the performance/dependability model for which decisions have to be made. Additionally, we may also use different stages to model time dependent changes in the decision process.

The general fluid model defined by (1) or (3) is not easy to handle, even a simulation requires some effort. Therefore, we describe the decision process by a continuous time Markov chain with a finite state space. This implies that discretization is used and an approximation error is introduced. However, this approximation error is usually much smaller than the modeling error which results from describing the complex human decision process by a fairly simple stochastic model.

First, we shift the state space from (θ_{-1}, θ_1) to $(0, \theta_1 - \theta_{-1})$. Then for $n_d > 2$ define $\Delta = (\theta_1 - \theta_{-1})/(n_d - 2)$. Now define an absorbing Markov chain with state space $\mathcal{S}_d = \{0, \ldots, n_d - 1\}$. The states 0 and $n_d - 1$ are absorbing and the remaining states are transient. As long as the stage remains constant we have $E(x(h)|x(0)) = x(0) + h\mu_k + o(h)$ and $Var(x(h)|x(0)) = h\sigma_k^2 + o(h)$. Both conditions are sufficient for a Markov chain to converge in the limit (i.e., $\Delta \to 0$) towards the fluid model [3]. Only at times when the decision stage changes, the relation does no longer hold. However, if such changes occur rarely (i.e., with probability 1 finitely often in each finite interval), then the following approximation still gives good results if n_d is chosen adequately.

For the Markov chain the initial distribution is defined as

$$\varpi(i) = \int_{(i-1)\Delta}^{i\Delta} \pi(y)dy. \tag{4}$$

Obviously ϖ defines a probability distribution on $\mathcal{S}_d$ where the probability of being initially in one of the absorbing states 0 or $n_d - 1$ is zero. Similarly, initial distributions ϖ_{-1} and ϖ_1 can be computed. If the initial distribution depends on the decision stage we write $\pi^{(k)}$ and $\varpi^{(k)}$.

Now the following transition probabilities are defined [3]

$$q_{i,j}^{(k)} = \begin{cases} \frac{\sigma_k^2}{2\Delta^2} + \frac{\mu_k^-}{\Delta} & \text{if } i = j - 1, \\ \frac{\sigma_k^2}{2\Delta^2} + \frac{\mu_k^+}{\Delta} & \text{if } i = j + 1, \\ -\frac{\sigma_k^2}{\Delta^2} - \frac{|\mu_k|}{\Delta} & \text{if } i = j, \\ 0 & \text{otherwise,} \end{cases} \tag{5}$$

where $\mu_k^- = \max(0, -\mu_k)$ and $\mu_k^+ = \max(0, \mu_k)$. The rates define a birth death process with absorbing boundaries at 0 and $n_d - 1$. For later use we define the following matrices for decisions in stage k. $\boldsymbol{Q}_0^{(k)}$ as a $n_d - 2 \times n_d - 2$ matrix that contains all transitions between states from $\{1, \ldots, n_d - 2\}$ plus the diagonal elements, i.e. all transient states, $\boldsymbol{q}_{-1}^{(k)}$ is a vector of length $n_d - 2$ with $\boldsymbol{q}_{-1}^{(k)}(1) = q_{1,0}^{(k)}$ and $\boldsymbol{q}_{-1}^{(k)}(i) = 0$ for $i \neq 1$, the transition in absorbing state 0 (i.e., decision -1) and $\boldsymbol{q}_1^{(k)}$ is a vector of length $n_d - 2$ with $\boldsymbol{q}_{-1}^{(k)}(n_d - 2) = q_{n_d-2, n_d-1}^{(k)}$ and $\boldsymbol{q}_{-1}^{(k)}(i) = 0$ for $i \neq n_d - 2$, the transition in absorbing state $n_d - 1$ (i.e., decision 1).

With the matrices decisions can be analyzed. The probability of making decision $d \in \{-1, 1\}$ in stage k is given by

$$pd_d = \varpi \left(-\boldsymbol{Q}_0^{(k)}\right)^{-1} \boldsymbol{q}_d^{(k)}. \tag{6}$$

The probability of choosing decision d until time t in stage k under the condition that decision d is chosen is given by

$$F_d(t) = \frac{1}{pd_d} \varpi \int_0^t e^{\tau \boldsymbol{Q}_0^{(k)}} \boldsymbol{q}_d^{(k)} d\tau. \tag{7}$$

4   Combined Models

A wide variety of different approaches for modeling systems according to performance and dependability exists. We consider here mainly models that are mapped on Markov chains and can then be analyzed using numerical techniques or, if the state space becomes too large, simulation. In the following we first describe the basic models and introduce afterwards the integration of models for human decisions.

4.1   Markovian Performance and Dependability Models

In Markov modeling for performance and dependability queuing networks, stochastic Petri nets or stochastic process algebras [1, 9, 22] are often applied. We do not introduce the different approaches in detail because this is common knowledge in the performance modeling area. Instead we describe a more abstract approach in the flavor of stochastic automata networks [14] which allows compositional modeling. We use the name performance model here also for dependability models since both types of models are specified with similar formalisms.

We consider a finite state system and assume that the performance model without the model for human decisions can be described by a continuous time Markov chain with state space $\mathcal{S}_p = \{0, \ldots, n_p - 1\}$. Let $\boldsymbol{Q}_\tau$ be a $n_p \times n_p$ generator matrix which contains all transition rates that do not depend on human decisions. Since $\boldsymbol{Q}_\tau$ is a generator matrix, the row sum is 0, non-diagonal elements are non-negative and diagonal elements are non-positive.

Human decisions, if made, immediately have an influence of the performance model. This means that with the decision the state of the performance model changes instantaneously. Since we have only two choices, -1 or 1, only two events can occur. Decisions are made by the human decision maker and are driven by the decision model, the performance model only reacts. Let R_{-1} and R_1 be two stochastic $n_p \times n_p$ matrices. $R_d(i, j)$ with $d \in \{-1, 1\}$ includes the probability that the state of the performance model changes from i to j when the human decision maker makes decision d in state i. If the decision has no influence on the current state, then $R_d(i, i) = 1$, the state remains. Matrices R_{-1} and R_1 specify the interface of the technical system to the human decision maker.

State changes due to decisions may for example describe the start of a repair (i.e., the state changes from defect to repair), the adding of a new server (i.e., the number of available servers is increased by 1), or the selection of a destination for an arriving task (i.e., the population in the destination is increased by 1). In all cases state space generation for the performance model has to consider all possible sequence of decisions by the human operator.

A state $i \in S_p$ is potentially reachable from a state $j \in S_p$, if a path in the matrix $Q_\tau + R_1 + R_{-1}$ exists. The state space of the performance model is potentially irreducible if every state is potentially reachable from every other state.

4.2   Integration of Human Decision Models

If a performance model and a model for the human decisions are available, then both have to be combined to build a composed Markov chain. Let (Q_τ, R_{-1}, R_1) be the matrix description of the performance model including the interface to the decision model. $\left(\varpi_{-1}, \varpi_1, Q_0^{(k)}, q_{-1}^{(k)}, q_1^{(k)}\right)_{k=1,\ldots,K}$ describes the decision model for K different stages. The selection of a stage depends in our model on the state of the performance model as it can be observed by the decision maker.

We first introduce the case that decision stages are directly related to subsets of the state space. Let $S_p^{(k)}$ ($k = 1, \ldots, K$) be a partition of the state space S_p (i.e., $S_p^{(k)} \cap S_p^{(l)} = \emptyset$ for $k \neq l$ and $\cup_{k=1}^{K} S_p^{(k)} = S_p$). Define $n_p \times n_p$ indicator matrices $I^{(k)}$ such that

$$I^{(k)}(i, j) = \begin{cases} 1 \text{ if } i = j \wedge i \in S^{(k)}, \\ 0 \text{ otherwise.} \end{cases} \tag{8}$$

With these indicator matrices, the generator matrix of the Markov chain for the composed model can be built.

$$Q = \underbrace{I_{n_d} \otimes Q_\tau + \left(\sum_{k=1}^{K} I^{(k)} Q_0^{(k)}\right) \otimes I_{n_p}}_{A}$$

$$+ \underbrace{\left(\sum_{k=1}^{K} I^{(k)} q_1^{(k)} \varpi_1\right) \otimes R_1}_{B} + \underbrace{\left(\sum_{k=1}^{K} I^{(k)} q_{-1}^{(k)} \varpi_{-1}\right) \otimes R_{-1}}_{C} \tag{9}$$

where I_n is the identity matrix of order n and $\otimes$ is the Kronecker product. If the initial distribution does not depend on the decision we have $\varpi_1 = \varpi_{-1} = \varpi$.

Q is a $n_p n_d \times n_p n_d$ matrix which can be represented by the sum of the matrices $A + B + C$ which have a very specific structure. All matrices can be structured into $n_d \times n_d$ blocks of size $n_p \times n_p$. Matrix A describes a finite quasi birth-death process

$$
\begin{pmatrix}
A_{1,1} & A_{1,2} & 0 & \cdots & & \cdots & & 0 \\
A_{2,1} & A_{2,2} & A_2 & \ddots & & \ddots & & \vdots \\
0 & \ddots & \ddots & & \ddots & & \cdots & 0 \\
\vdots & \ddots & \ddots & & \ddots & & \cdots & \vdots \\
\vdots & & \ddots & \ddots & A_{n_d-1,n_d-2} & A_{n_d-1,n_d-1} & A_{n_d-1,n_d} \\
0 & \cdots & \cdots & & 0 & A_{n_d,n_d-1} & A_{n_d,n_d}
\end{pmatrix}
$$

where $A_{i,i} = Q_\tau + \sum_{k=1}^K I_k Q_0^{(k)}(i,i)$ and $A_{i,j} = \sum_{k=1}^K I_k Q_0^{(k)}(i,j)$ with $j \in \{i-1, i+1\}$. Furthermore

$$
B = \begin{pmatrix}
0 & \cdots\cdots & 0 \\
\vdots & \ddots\ddots & \vdots \\
0 & \cdots\cdots & 0 \\
B_1 & \cdots\cdots & B_{n_d}
\end{pmatrix}
\quad \text{and} \quad
C = \begin{pmatrix}
C_1 & \cdots\cdots & C_{n_d} \\
0 & \cdots\cdots & 0 \\
\vdots & \ddots\ddots & \vdots \\
0 & \cdots\cdots & 0
\end{pmatrix}.
$$

where $B_i = \varpi_1(i)\left(\sum_{k=1}^K I^{(k)} q_{n_d,n_d+1}^{(k)}\right) R_1$ and $C_i = \varpi_{-1}(i)\left(\sum_{k=1}^K I^{(k)} q_{1,0}^{(k)}\right) R_{-1}$. In particular if the vectors ϖ_{-1} and ϖ_1 are sparse, the matrix has a very sparse structure which can be exploited in solution techniques. If $\sigma_k > 0$ for all k (i.e., all decisions include some uncertainty), then Q is irreducible if and only if the performance model is potentially irreducible.

The above description assumes that the human operator can identify subsets of states of the performance model exactly. This is not always possible. E.g., it is usually not possible to completely determine the internal state of a machine to decide whether a maintenance is necessary or not. Often the internal state can only be estimated and the decision stage depends on the estimate. Let $s_i(k)$ be the probability that the decision maker guesses stage k under the condition that the performance model is in state i. The values form a probability distribution over the set of decisions stages, i.e., $\sum_{k=1}^K s_i(k) = 1$ for all $i \in S_p$. Now define diagonal matrices $S^{(k)}$ by

$$
S^{(k)}(i,j) = \begin{cases} s_i(k) & \text{if } i = j, \\ 0 & \text{otherwise.} \end{cases} \tag{10}
$$

If matrices $S^{(k)}$ rather than $I^{(k)}$ are used in (9), then the matrix of the composed models under uncertainty of the decision stage is computed.

4.3   Model Analysis

The model can be analyzed with respect to stationary or transient measures. In both cases rewards are usually defined to compute results like throughputs, sojourn times or the availability and reliability of a system. Let $r \in \mathbb{R}^{n_p}$ be a reward vector that assigns rewards to the states of the performance model.

Analysis can be performed numerically or by simulation, if the state space is too large for a numerical analysis. We consider here only the numerical analysis. For stationary analysis the system of linear equations

$$pQ = 0 \text{ w.r.t. } p\mathbb{1} = 1 \tag{11}$$

has to be solved. The expected stationary reward is then given $E_s = p\left(I_{n_p} \otimes \mathbb{1}_{n_d}\right)r$. The specific structure of matrix Q supports the use of block iterative methods [2,21].

For transient analysis the linear differential equations

$$p_t = p_0 e^{tQ} = e^{-\alpha t} p_0 \sum_{h=0}^{\infty} \frac{(\alpha t)^h}{h!} \left(Q/\alpha + I\right)^h \tag{12}$$

with initial distribution p_0 and $\alpha \geq \max_{i \in \mathcal{S}}\left(|Q(i,i)|\right)$ have to be solved. The instantaneous reward at time t is then given by $E_t = p_t\left(I_{n_p} \otimes \mathbb{1}_{n_d}\right)r$ and the accumulated reward in the interval $[0,t]$ equals $E_{at} = \int_0^t p_\tau\left(I_{n_p} \otimes \mathbb{1}_{n_d}\right)r d\tau$.

5   Examples

We consider small examples to show the modeling capabilities of the proposed approach.

5.1   A Simple Routing Selection

The first model is a simple open queueing model with two finite capacity queues with capacity 10. The arrival rate of new items is λ and queues have a mean service time of $\omega^{-1} = 1$. Items may become impatient during their waiting time according to an exponential distribution with rate $\nu = 0.2$ per waiting item. Impatient items leave the system without being served, items trying to enter a queue that is full get lost. If a new item arrives, it is first stored in a router and a human decision maker decides whether to put it in the first or second queue. We assume that the queue state is not exactly observable. Decision making can be modeled by a diffusion model (3) with drift rate $\mu(n_1, n_2) = 10(n_2 - n_1)$ where n_1 is the current population in queue 1 and n_2 is the current population in queue 2 and constant variance $\sigma_k = 10$. Since $n_1 - n_2 \in [-10, 10]$, we have $K = 21$ different non-zero drift rates. If the decision maker decides for 1, then the item is put into queue 1, for decision -1 it is put in queue 2. After a decision the decision maker waits for the next item. If a new item arrives and the decision for the previous one has not been made, then the item selects randomly one of

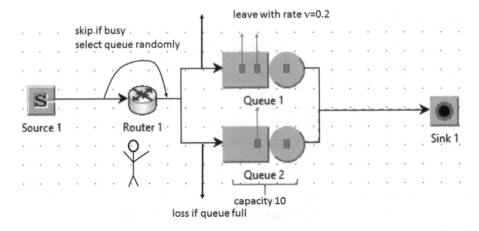

Fig. 1. First example model

the two queues. The model is shown in Fig. 1. The measure which is analyzed is the throughput of successfully served items in steady state.

We assume that $\theta_1 = \theta$ and $\theta_{-1} = -\theta$ and consider $\theta = \frac{1}{2}, \frac{3}{2}, \frac{5}{2}, \frac{7}{2}, \frac{9}{2}$ and values $\Delta = 0.5, 1$ for discretization. A smaller value for θ puts more pressure on the decision maker which means that decisions are made faster but with a higher probability of a wrong decision. If Poisson arrivals and exponential service time distributions are assumed, then the number of states of the model equals $121(2\theta/\Delta + 2)$ if $\varpi_{-1} \neq \varpi_1$ and $121(2\theta/\Delta + 1)$ for $\varpi_{-1} = \varpi_1$. We first assume that the decision maker starts always in $x(0) = 0$ independently of the last decision.

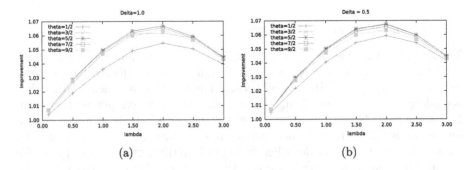

Fig. 2. Relative improvement of the throughput for the model with Markovian arrivals and services.

In Fig. 2 the improvement of using a decision maker rather than a random selection of queues is shown for the different configurations. It can be noticed that the choice of Δ has only a minor effect on the results which implies that

$\Delta = 0.5$ is sufficient to approximate the fluid model for decision making. If we compare the results for different values of θ, it can be noticed that $\theta = \frac{1}{2}$ is too small, the decision maker decides too fast and the error probability is too large. For the other values of θ, the differences are small. In general $\theta = \frac{5}{2}$ gives the best results over all analyzed arrival rates. We also analyzed the system for $\theta = \frac{5}{2}$ and decision dependent starting points. I.e., if the last decision was 1 the decision maker starts with $x(0) = 1$ or 2 and if the decision was -1 the starting point is $x(0) = -1$ or -2. However, in the case results are worse than before. The results are shown on the left side of Fig. 4.

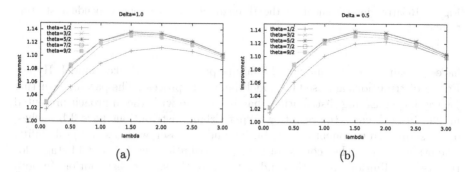

Fig. 3. Relative improvement of the throughput for the model with Erlang 3 inter-arrival times and $H2$ service times.

We consider a different version of the model where inter-arrival times are Erlang 3 distributed and the service times have a 2 phase Hyper-Exponential distribution with $CV = 10$. Two effects can be expected: First, the time between arrivals becomes more predictable and the decision maker will lose fewer items due to early arrivals. Second, the difference in the queue length will become larger due to a high variability in service times. Thus, we can assume larger improvements compared to a random selection of the queue. This effect can indeed be observed in Fig. 3 which shows the results for the example. The improvement is twice as large as before but the shape of the curves is similar. We also analyzed the model with a decision maker who remembers the last decision and starts near the corresponding boundary. The corresponding results are shown on the right side of Fig. 4. In this version decision dependent starting points result in a minor improvement of the throughput.

5.2   A Maintenance Model

As a second example we consider a simple maintenance model consisting of two machines and a single repairman. Both machines have Weibull distributed failure time with shape parameter 3.0, scale parameter 11.199 for machine 1 and shape parameter 2.0, scale parameter 11.285 for machine 2. Both Weibull distributions

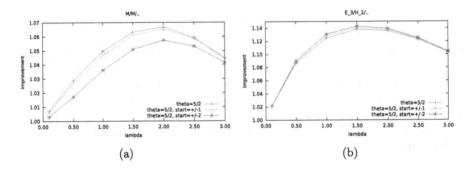

Fig. 4. Relative improvement of the throughput with decision dependent starting points

have a mean of 10. To analyze the failure process in a Markov model, Hyper-Erlang distributions are used to model the failure process. The parameter fitting for the Hyper-Erlang distributions has been done with the approach presented in [23]. In both cases the resulting Hyper-Erlang distributions have 3 branches, one describes an exponential distribution and is chosen with a small probability. The other two branches consist of Erlang distributions with 5 and 14 stages for the first distribution and with 3 and 6 stages for the second distribution. In both cases, the differences between the Weibull distributions and the Hyper-Erlang approximations are very small.

A single repairman inspects the machines cyclically. If a machine fails, the repairman starts immediately a repair, if he is not involved in the repair or maintenance of the other machine. The duration of a repair operation is Erlang 3 distributed with mean 2. During inspection the repairman decides whether to continue or to do a maintenance operation. A maintenance operation has a Erlang 3 distribution duration with mean 0.5. After a repair or maintenance operation, the machine is again in its initial state which means that the Hyper-Erlang distribution starts again.

The decision process of the repairman is driven by a diffusion process with parameters μ and σ. Parameter μ depends on the state of the corresponding failure distribution. If a Hyper-Erlang distribution is in an Erlang branch with r phases, then $\mu = 1$ if the phase number is larger or equal $\lceil r \rceil$ and $\mu = -1$ otherwise. Parameter σ is always 2. This reflects the situation that the time to failure is not exactly known but can estimated by observing a machine. We assume that a decision to do a maintenance operation is taken if the decision process reaches threshold θ, if threshold $-\theta$ is reached, then the machine keeps running and the repairman inspects the other machine.

The goal is to analyze the availability of the machine (i.e., the time when machines are running). Figure 5 shows the availability for different values of the threshold θ. It can be seen that the largest availability is achieved by choosing θ slightly smaller than 2.

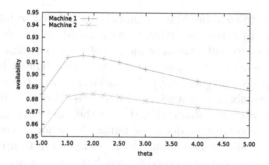

Fig. 5. Availability of the machines depending on the threshold θ.

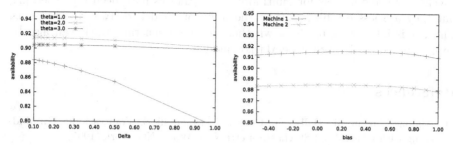

(a) Influence of the discretization parameter Δ on the result.

(b) Influence of the starting point on the availability for $\theta = 2$.

Fig. 6. Influence of discretization parameter and starting point on the result.

Figure 6a shows the influence of the discretization parameter Δ on the results for 3 different values of θ. It can be noticed that for smaller values of θ the choice of a small Δ is more important than for larger values. Beyond $\Delta = 0.15$, the results remain almost identical if Δ is further reduced. Another important question is, whether the availability can be increased if the decision process starts with some bias, i.e., not at 0. Figure 6b shows the results for $\theta = 2$, when the decision process starts initially at some point between $[-0.5, 1]$. A positive bias results in shorter decision time if a maintenance is chosen and a negative bias results in shorter decision times when the repairman continues inspection. It can be seen that the influence of the starting point on the availability is marginal as long as the starting point is not chosen near to one of the boundaries such that random effects dominate the related decision.

6   Conclusions

The paper presents an approach to integrate models for human decision making known from psychology into commonly used models for performance and dependability. Using continuous time Markov chains as the joint model type, known approaches to combine Markov models are applied resulting in an overall Markov model that can be analyzed with standard means. The proposed approach allows one to build coherent models of socio-technical systems.

The paper presents a first approach in integrating human behavior in performance and dependability modeling. It considers mainly the technical aspects of how to build, combine and analyze the models. However, from a practical point of view the correct parameterization of the models is extremely important. Especially the correct modeling of human behavior in complex technical environments is a challenge and should be investigated in future research.

Another restriction of the proposed approach is that we consider only two way decisions for human operators. In practice often more than two choices exist. In psychology some models for multi-alternative choices have been published [24] but they are less established than the diffusion model for two-way decisions. Thus, it is interesting to check whether the more general models can also be approximated by discrete state Markov models.

References

1. Ajmone-Marsan, M., Balbo, G., Conte, G., Donatelli, S., Franceschinis, G.: Modelling with Generalized Stochastic Petri Nets. Wiley, New York (1995)
2. Buchholz, P., Dayar, T.: Block SOR for Kronecker structured representations. Linear Algebra Appl. **386**, 83–109 (2004)
3. Cerrato, M., Lo, C.C., Skindilias, K.: Adaptive continuous time Markov chain approximation model to general jump-diffusions. Working Paper 2011–16, University of Clasgow, Business School of Economics (2011)
4. Curram, S.: Representing intelligent decision making in discrete event simulation: a stochastic neural network approach. Ph.D. thesis, University of Warwick (1997)
5. Dhillon, B.S.: Human Reliability, Error, and Human Factors in Power Generation. Springer, Heidelberg (2014)
6. Diederich, A.: Simple matrix methods for analyzing diffusion models of choice probability, choice response time, and simple response time. J. Math. Psychol. **47**, 304–322 (2003)
7. Diederich, A.: A multi-stage attention-switching model account for payoff effects on perceptual decision taks with manipulated prcoessing order. Decision **3**(2), 81–114 (2016)
8. Flitman, A.M., Hurrion, R.D.: Linking discrete-event simulation models with expert systems. J. Oper. Res. Soc. **38**(8), 723–733 (1987)
9. Hillston, J.: A compositional approach for performance modelling. Ph.D. thesis, University of Edinburgh, Department of Computer Science (1994)
10. Lee, S., Son, Y.: Integrated human decision making model under belief-desire-intention framework for crowd simulation. In: Proceedings of the Winter Simulation Conference, pp. 886–894 (2008)
11. Lee, S., Son, Y., Jin, J.: An integrated human decision making model for evacuation scenarios under a BDI framework. ACM Trans. Model. Comput. Simul. **20**(4), 23:1–23:24 (2010)
12. Lyu, J., Gunasekaran, A.: An intelligent simulation model to evaluate scheduling strategies in a steel company. Int. J. Syst. Sci. **28**(6), 611–616 (1997)

13. Norling, E., Sonenberg, L., Rönnquist, R.: Enhancing multi-agent based simulation with human-like decision making strategies. In: Moss, S., Davidsson, P. (eds.) MABS 2000. LNCS (LNAI), vol. 1979, pp. 214–228. Springer, Heidelberg (2001)

14. Plateau, B., Fourneau, J.M.: A methodology for solving Markov models of parallel systems. J. Parallel Distrib. Comput. **12**, 370–387 (1991)

15. Ratcliff, R.: A theory of memory retrieval. Psychol. Rev. **85**, 59–108 (1978)

16. Ratcliff, R., Smith, P.L., Brown, S.D., McKoon, G.: Diffusion decision model: current issues and history. Trends Cogn. Sci. **20**(4), 260–281 (2016)

17. Robinson, S.: Modeling human interaction in organizational systems. In: Fishwick, P. (ed.) Handbook of Dynamic System Modeling. CRC Press, Boca Raton (2007)

18. Robinson, S., Edwards, J.S., Yongfa, W.: An expert systems approach to simulating the human decision maker. In: Proceedings of the Winter Simulation Conference (1998)

19. Shen, W., Maturana, F., Norrie, D.H.: MetaMorph II: an agent-based architecture for distributed intelligent design and manufacturing. J. Intell. Manuf. **11**(3), 237–251 (2000)

20. Sridharan, V., Mohanavadivu, P.: Reliability and availability analysis for two non-identical unit parallel systems with common cause failures and human errors. Microelectron. Reliab. **37**(5), 747–752 (1997)

21. Stewart, W.J.: Introduction to the Numerical Solution of Markov chains. Princeton University Press, Princeton (1994)

22. Stewart, W.J.: Probability, Markov Chains, Queues, and Simulation. Princeton University Press, Princeton (2009)

23. Thümmler, A., Buchholz, P., Telek, M.: A novel approach for phase-type fitting with the EM algorithm. IEEE Trans. Dependable Secur. Comput. **3**(3), 245–258 (2006)

24. Tsetsos, K., Usher, M., McCelland, J.L.: Testing multi-alternative decision models with non-stationary evidence. Front. Neurosci. **5**, 63 (2011)

25. Tuerlinckx, F., Maris, E., Ratcliff, R., Boeck, P.D.: A comparison of four methods for simulating the diffusion process. Behav. Res. Methods Instrum. Comput. **33**(4), 443–456 (2001)

26. van Dam, K.H.: Capturing socio-technical systems with agent-based modelling. Ph.D. thesis, Technology, Policy and Management, TU Delft (2009)

27. Williams, T.: Simulating the man-in-the-loop. OR Insight **4**(9), 17–21 (1996)

28. Zhao, X., Venkateswaran, J., Son, Y.-J.: Modeling human operator decision-making in manufacturing systems using BDI agent paradigm. In: IIE Annual Conference and Exposition (2005)

29. Zülch, G.: Modelling and simulation of human decision-making in manufacturing systems. In: Proceedings of the Winter Simulation Conference, pp. 947–953 (2006)

Combining Simulation and Mean Field Analysis in Quantitative Evaluation of Crowd Evacuation Scenarios

Sandro Mehic[1]([✉]), Kumiko Tadano[1,2], and Enrico Vicario[1]

[1] Department of Information Engineering, University of Florence, Florence, Italy
{sandro.mehic,enrico.vicario}@unifi.it
[2] NEC Corporation, Tokyo, Japan
k-tadano@bq.jp.nec.com

Abstract. Crowd movement analysis methods structurally suffer from the problems of scalability and lack of empirical data. Macro scale approaches can tackle larger crowds, but they prevent direct representation of dynamics occurring on the micro scale which may be more easily related to observations.

In this work, we present and experiment a hierarchical approach which aims at conciliating the contrast through the combination of fine grained simulation and coarse grained analytical techniques. To this end, agent based simulation is applied on micro scale patches where mechanisms resembling reality can be more easily reproduced. Measurements on simulation results are then used as parameters for an analytical model which exploits mean field techniques to efficiently approximate the emerging crowd behavior through the fluid limit obtained when the crowd density tends to infinity.

The approach is experimented with reference to a crowd evacuation scenario, using NetLogo as engine for spatial agent based simulation and JSam framework as a solution engine for mean field analysis.

1 Introduction

Motion of human crowds is a complex phenomenon resulting from the combination of several and different factors such as goals and motion strategies of individuals, their social and emotional attitudes, geometric characteristics of the context, availability of information.

Crowd modelling, including the problems of evacuation and steering, have always been a popular field of research, which is now gaining further interest from advances in computational power and methods and from the growing potential social and security utility, e.g. in the prevention of casualties during evacuation, large scale events, customer flow guidance, etc. With greater availability of data, higher frequency of social events as well as their increasing size, it is becoming more important to be able to accurately predict the crowd movement in order to be able to respond to emergency situations. Being able to anticipate bottlenecks

© Springer International Publishing AG 2016
D. Fiems et al. (Eds.): EPEW 2016, LNCS 9951, pp. 174–186, 2016.
DOI: 10.1007/978-3-319-46433-6_12

and critical points in crowd dynamics on a large scale can help safeguard higher number of lives in case of a disaster, whether natural or man-made.

Most works in the literature address small scale and localized problems, like evacuation from a single structure, such as in [1] where emergency evacuation from a single building is analysed, or in [2] where impact of a social abstraction layer is studied during an evacuation from a metro station in London, or in [3] where more relevance is given to simulating the phenomenon of evacuation panic.

In various works, parameters of crowd behaviour are derived from video footage or other information sources, enabling extraction of significant features such as statistics of groups and their impact on motion parameters, [4] and [5]. Computer vision methods can be applied to extract specific features that can later be used to model or predict events in crowd movements, such as [6] where the importance of stop-and-go waves is studied from the footage of Hajj pilgrimage. As a common trait, these approaches rely on empirical data referring to specific scenarios, actually occurred and observed in reality, and provide only partial insight for prospective studies addressing behaviour in potential different contexts.

Aside from the difficulty to define models that are highly scalable caused by the computational complexity of the task, the major obstacle is the lack of empirical data, and the difficulty to conduct experiments on such a scale.

Works addressing movements of crowds with high population size, mostly rely on models based on the fluid dynamic approach, such as in [7] and [8]. In this case, the basic idea enabling the study of larger crowds revolves around aggregating single agents in larger groups that behave as a single entity. This approach can improve significantly the performance, but is still limited as the computational cost still depends on the number of agents. So, this class of models may be able to afford problems with tens of thousands of agents, but still suffer when tending towards crowd movements addressed in modern day metropolis, which may count several million people.

To address the issues of scale and poor integration between different scale solutions we propose a technique that combines in a hierarchical way different methods to make the best of each solution. We propose a technique to elaborate the parameters from micro scale simulations, that can be successfully used to model a macro scale model with parallel Markov chains. In addition we show how mean field theory can be applied to a large number of parallel Markov chains to effectively remove the dependency of the solution technique on the number of agents. This way large scale crowd movement scenarios of potentially infinite number of agents can be solved.

2   Approach

Hierarchical approach proposed in this paper uses three different layers of scale in order to achieve an analytic solution on a macro scale level by using the mean field approximation, while using simulation on micro scale for parameter estimation.

As a first layer we use fine grained, i.e. microscopic, agent base modelling to obtain parameters such as transition probabilities and sojourn times in zones of the map. The use of agent-based modelling allows for introduction of rich physical and social interaction between individuals, that can be implemented by already known methods, such as social force, rule based models, etc. Complex social behaviour paradigms can be introduced as they have been found to be important for a more realistic representation of the crowd, such as the altruist policy that was found to be missing in most of the works on crowd movement analyzed in [9], or the conformism strategy that introduces the confidence of single agents to find their way or 'go with the flow', as studied in [10] and [11]. The movement strategy, also called the physical aspect of the agent, consist in searching for a nearest exit and orienting towards it. The physical and the social aspect together determines the movement of each single agent, in function of the context in which he finds himself. With simple strategies like these, a rich and complex behaviour is obtained, that reflects realistically the behaviour of crowd in real world environment.

The second layer of our proposal, i.e. mesoscopic, abstracts the single agent from the agent based simulation (ABS) in a discrete time Markov chain that is modelled using the parameters obtained from ABS. This part acts as coagulant between micro and macro scale, by observing individual agents and measuring their transition probabilities between regions of the map, and the relative sojourn times, we can extract through statistical means a more general representation of individuals movement on the map. State transitions are modelled in function of the fractions of other chains in various possible states, through which we maintain the agents interactions at macro scale. Each DTMC represents a single agent in a single region of the space that we intend to study, and the rich behaviour observed in ABS is extracted as the probability functions for state transitions of the single chain.

The third and the final layer of our proposal, i.e. macroscopic, consist in the composition of N parallel Markov chains, where N is the number of agents in the system, obtained at the second step and the composition of M adjacent regions that represent the physical space, in our example a city. The large-scale physical space is at this point represented as a graph, with single regions as nodes and the connectivity between them as edges of the graph.

The most important step of this final layer is the mean field approximation, such as presented in [12] and [13]. While other works usually fail to analyze systems with large number of agents, with the application of the mean field approximation we can find the solution by solving a set of ordinary difference equations, that is computationally dependent on the number of states, and not on the number of agents. This way we are able to solve crowd movement models with arbitrary number of agent, potentially infinite, whose solution is exact in probability. Observing the results from the work on the mean field Markov chains [12], the approximation of the analytical solution is precise in probability, i.e. for $N \to \infty$, meaning that higher the number of agents, more precise the solution. By effectively using this theoretical result we can be sure that our solution is more precise, the higher the number of agents, and hence we convert one of the main issues of other techniques in a strength.

The problem of the computational cost is relieved by the introduction of the mean field approximation, and the dependency on the number of agent is removed at this scale level. We introduce a Markov chain that acts as an agent with the number of states equal to the product of kernel of each abstraction layer, in our experiments only the physical aspect, so the states of the Markov chain represent the regions of the map. In our proposal the state space is held relatively low by using the hierarchical approach and defining states on a high enough level.

3 Methods

3.1 Agent Based Modelling

In this work we have developed our own agent based model to simulate the crowd movement in a well defined region. The model was developed in *NetLogo* tool, and is used to estimate the parameters for the construction of Markov chains for the higher levels.

The final result is influenced by how these fine grained models are implemented, however if there is necessity to represent different local behaviour, different agent based models can be integrated, as long as they allow the calculation of transition probabilities and sojourn times for agents moving in the studied region.

Physical Aspect. In our model the physical aspect is resolved by discretisation of space and time. While there are more sophisticated models, using continuous space and time, like in [14], the main objective of this paper is on the higher level of abstraction and scalability of the approach.

The region is divided in cells, where each cell can be occupied by at most n agents. The size of a single cell should be estimated as to represent truthfully the dimensions of the space being modelled with respect to the size of the agents. In our case the spatial representation and the number of patches used to model roads and intersections was such that $n = 5$ was a realistic estimate for the maximal occupancy of a single patch.

The speed of each agent is modelled as a geometric variable when no obstacles are present, so after actually introducing a single agent in a context with others, as well as with their social strategies, this behaviour becomes more complex.

Spatial Model. The *NetLogo* model was divided in two functional parts, first for the generation of the map and the environment of the simulation, while the second contained the agents logic for moving in the generated context. For our crowd movement models we have chosen to represent the space as a graph, where the intersections are abstracted as nodes and the streets between them as edges. In this way the region that should be simulated can be formalized in an external document, such as XML or other, and we can use additional tools when needed for set operations on various regions, when a different dimension of the region is preferred. Thus a map is represented as a graph, where nodes have the additional field, such as the width, height, radius of intersection, and each of them can have different behaviours, primarily (1) entry points where new agents enter from adjacent regions or buildings, (2) exit points where agents exit the region or reach their destination, (3) interest points that can be a part of a single agent behaviour that is determined to visit them, or just regular intersection without additional functionality. Edges in the graph represent the streets and aside from having the source and destination node, i.e. the places they connect, they are defined by their width and weight, where the later is used for calculating the optimal path for agents.

In our agent based model at each time step every single agent decides the best direction to follow in function of the physical and social reality of the context. Each agent has a list of destination nodes that he wishes to reach, and when his destination list is empty he will try to reach the nearest exit. This special case is useful for evacuation simulation. The movement of the agent is recalculated on the graph each time a node is reached. When an agent reaches a node, a weighted path to his destination is calculated, but only the next node to reach is given to the agent, in this way a dynamic update of weights is possible, through which we can model in more detail the formation of crowds in various scenarios.

The conformism in the model is introduced through a local variable *patience* of the agents, that represents the amount of time an agent is ready to wait in order to reach his destination following the current road. Each time an agent is unable to move towards his destination, his patience is decreased, and once it hits zero, the agent flags the current street as undesirable and will try to reach his destination by following different paths. In this way we are able to construct a rich behaviour on a regional scale, and resolve the blocking between agents.

The implementation described here is a simple agent based model, needed in order to estimate the parameters for a higher level abstraction of parallel discrete time Markov chains, but can be replaced by other simulation models as long as they allow the calculation of the needed parameters.

3.2   Mean Field Approximation and DTMC

Starting from the complex structure of a region in agent based modelling we can abstract to an upper hierarchical level, i.e. mesoscopic scale, where a single agent is represented as a discrete time Markov chain. In order to construct the DTMC, we divide the region in smaller subregions, and we model with the states the physical location of the agent, while the transitions represent the movement from single subregion to an adjacent one. An example can be seen in Fig. 1, where region is divided in 7 subregions.

In order to define a DTMC we need transition probabilities that truthfully model the change of physical location for the individual agent, and also those probabilities need to reflect the sojourn time of each agent.

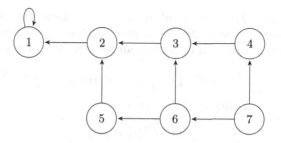

Fig. 1. Example of a region abstraction in a DMTC with 7 states representing subregions

Mean Field Approximation. As seen through parameter estimation from the micro scale a single agent can be abstracted to a DTMC, but we actually have many agents acting at the same time. The approach is to have N parallel interacting discrete time Markov chains and use the mean field approximation result from [12], from which we introduce the definition without memory, i.e. respectful to the Markov condition.

Let N be the number of interacting DTMCs all synchronized to the same time step. Each DTMC has a set of states $\epsilon = \{1, 2, \ldots S\}$ and with $X_n^N(t)$ we define the state of the n-th DTMC at time t.

Mean field approximation studies the behaviour of the system composed by N parallel Markov chains by analysing the occupancy vector $M^N(t)$ that is the vector of occupancy measures of all the chains. The occupancy measure of state i at time t is defined as:

$$M_i^N(t) = \frac{1}{N} \sum_{n=1}^{N} 1\{X_n^N(t) = i\}$$

where $1\{X_n^N(t) = i\}$ is equal to 1 if DTMC n is in the state i at time t and 0 otherwise.

The state transition of an individual DTMC is defined as a function of the current state of the DTMC and the occupancy vector of the system.

$$P\{X_n^N(t+1) = j \mid M^N(t) = \boldsymbol{m}, X_n^N(t) = i\} = K_{i,j}^N(\boldsymbol{m})$$

The value of $K_{i,j}^N(\boldsymbol{m})$ is calculated as a function of the occupancy measure of the state, and we have modelled it as a linear function whose slope and intercept are estimated from the micro scale simulation.

Self-loop transition from state i can occur if $K_{i,i}^N > 0$ and the values of K are constrained by

$$\sum_{j=1}^{S} K_{i,j}^N(\boldsymbol{m}) = 1$$

The occupancy vector $M^N(t)$ is a random variable and for different runs we obtain different results, but if we have that $N \to \infty$ we can use a deterministic approximation $\boldsymbol{\mu}(t)$. The main result of [12] is that under the assumption that the initial occupancy vector $M^N(0)$ converges almost surely to deterministic limit $\boldsymbol{\mu}(0)$, and the definition of $\boldsymbol{\mu}(t)$ as an iterative function starting from its initial value, i.e.

$$\boldsymbol{\mu}(t+1) = \boldsymbol{\mu}(t) \times K(\boldsymbol{m}(t))$$

Then for any fixed time t, almost surely:

$$\lim_{N \to \infty} M^N(t) = \boldsymbol{\mu}(t)$$

Following from this result, for relatively large numbers of parallel DTMCs we can use the deterministic approximation and the *fast simulation* algorithm provided in [12]. In this way the macro scale part of our model becomes independent of number of agents, i.e. population, that we want to study, effectively having the possibility to obtain results for systems with large number of people in real time, once the parameters for the DMTC are estimated from micro scale simulation.

3.3   Composition

As shown in Sect. 3.2, we can obtain a DTMC that represents a region of the space we are interested in studying, where each state models a subregion that are interconnected. This is done by encoding additional states that define the location of a single agents and transitions can model the change of physical aspect, i.e. movement from one region to another.

The separation of the physical space, for example a city map, in separate regions will determine the number of states in the final discrete time Markov chain.

The composition of N parallel DTMCs is the macro scale level of our approach, and thanks to the mean field approximation, the solution is obtained by solving a set of ordinary difference equations. At this scale we are no longer constrained by the number of agents in the model, but its computational cost is proportional to the number of states in the DTMC.

4   Experiments

In order to confront the mean field approximation on a macro-scale level after obtaining the parameters from agent based modelling, we have conducted a single monolithic simulation campaign, that serves as a ground truth, and have then decomposed the monolithic map in subregions that were then simulated independently in order to estimate the DTMCs needed for the macro-scale solution. In Sect. 4.1 we briefly explain the monolithic model while in the Sect. 4.2 we show different scenarios that are used to run the mean field approximation.

The map used for the experiments was generated starting from a residential block of the city of Florence, with the intention to include various types of map texture, such as dense intersections with small width of street that represent high density city zone, and low density with wide street regions that represent the large roads in modern city quarters.

Agents were uniformly distributed wrt the free space, hence leaving regions with higher density of streets more populated than others.

4.1 Monolithic Model

Monolithic model represents the map of the entire evacuation zone and can be seen in Fig. 2 represented as a graph that is used in our model to draw the topology of the evacuation site. The red node in the figure shows where the evacuation point is located, that all the agents are trying to reach, and which we assume to have infinite capacity.

The initial state is assumed to be 10000 agents distributed uniformly on the monolithic model, with all of them trying to reach the evacuation point. Our simulation campaign is based on multiple runs of the monolithic model and will serve as the ground truth for the hierarchical approach. In Fig. 3 we show the mean of the empirical cumulative distribution functions of the time units required for an evacuee to reach the exit destination. The grey area around the ECDF represents the mean plus and minus the standard deviation of the simulation campaign. We have collected data from 8 runs in order to have a more precise estimation of the mean evacuation times, and the simulation campaign took approximately *9 h:19 m* on a 8-core Intel Xeon Processor E5640 and 32 GB machine.

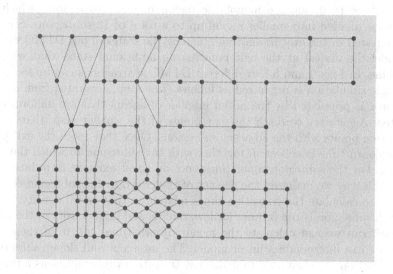

Fig. 2. The map of the monolithic model with the subdivision in 6 principal regions (Color figure online)

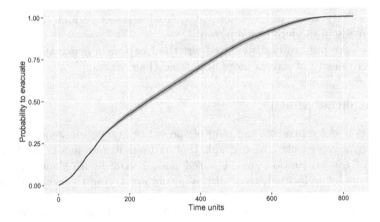

Fig. 3. The mean of the single empirical distribution functions of the probability to have evacuated by time t

4.2   Decomposed Region Scenarios

In order to be able to represent an agent as a DTMC we need the transition probabilities and the mean sojourn time in the state for that agent. The states of our DTMC represent the physical zones of the map, hence our approach is to divide the monolithic model in smaller regions that can be simulated independently from which we can estimate the needed parameters.

We chose to divide the monolithic region in 6 macro regions, which can further be divided into smaller region up to a total of 12 subregions. Since we are interested in the evacuation times, among the states of our DTMC we have to model the arrival at the exit point as an additional state, that we make absorbing. In Figs. 4 and 5 four different DTMC abstractions can be seen.

Single simulation is organized as follows. The map, generated from a graph structure, is populated by the initial number of agents that are uniformly distributed. Agents try to reach the exit points of the smaller map, that are the connection points with the adjacent subregions. Once they reach the exit point, their sojourn time is collected, together with the subregion to which the agent is going. For the subregion simulations once an agent exits we introduce a new agent, this way we maintain the density of the crowd in the subregion constant in order to calculate the transition times relative to that crowd density.

Each subregion is run 3 times for low, middle and high density of the crowd. For each run we can calculate the mean sojourn time, and from these three values we can interpolate a linear model. The intercept and slope values of the linear function are then used to define the transition probabilities as a function of the crowd density for a state that models the subregion of the DTMC. This way we define the transition probabilities in the mean field approximation as the function of the occupancy vector of states.

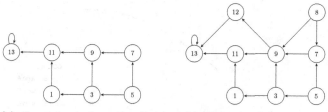

(a) Scenario A for map division into 7 DTMC states

(b) Scenario B for map division into 9 DTMC states

Fig. 4. DTMC abstractions from the monolithic model

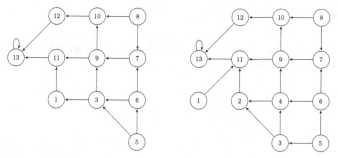

(a) Scenario C for map division into 11 DTMC states

(b) Scenario D for map division into 13 DTMC states

Fig. 5. DTMC abstractions from the monolithic model

5   Results

From the experiments conducted we were able to estimate the parameters for the mean field approximation models of four different scenarios, as shown in Figs. 4 and 5. The new models that are obtained can be solved as a set of ODEs and we have used the Jsam tool developed by the authors of [13].

In Fig. 6 we can see the confrontation between the probability to have reached the evacuation point by time t in the monolithic model and the mean field approximation obtained through regional decomposition. In Fig. 7 we show the error made by the mean field approximation model wrt the simulation from the monolithic model.

From the results we can observe that the mean field approximation models maintain the qualitative behaviour of the monolithic simulation, and the accuracy of the solution is different for the four scenarios. For scenario A with 6 subregions we see a larger error in the transitional phase of the evacuation simulation, while it approximates better the time all agents reach the evacuation state. On the other hand, scenario D commits the smallest error in the time interval going from 200 til 500 time units, but it commits larger error at 830 time units that was estimated as the evacuation time limit from the monolithic model.

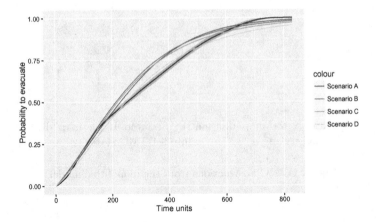

Fig. 6. The probability to have reached the evacuation point at time t, for four scenarios compared to the monolithic model

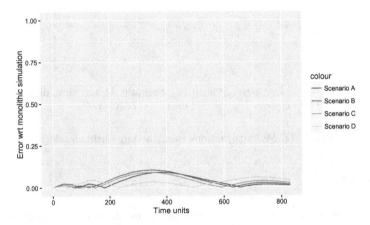

Fig. 7. Error of mean field approximation of different scenarios wrt the monolithic model

Table 1. Maximum and cumulative error in time interval [0,830]

	Max. error	Cumulative error
Scenario A	0.09175553	30.20908
Scenario B	0.1054587	36.38478
Scenario C	0.09103714	31.49491
Scenario D	0.06077186	24.61439

In Table 1 we show the maximum and the cumulative error made by different scenarios wrt the monolithic model simulation.

The hierarchical approach presented in this paper enables the decomposition of large crowd movement models into smaller models that can be easily resolved through simulation. Although the simulation times of the monolithic model and the decomposed subregions remain in the same order of magnitude, the modular nature of the decomposed model allows a more flexible crowd movement analysis through modification of only certain regions. The hierarchical approach allows for a quicker study of the impact of modification of urban areas on the mobility of the crowd. If we should study the impact on the mobility in function of the availability of certain areas, we only need to run the simulations of the modified areas, while the parameters estimated for the rest of the subregions remain the same. The cost of a calculating the solution is then considerably smaller wrt the monolithic solution.

Our proposal could prove to be crucial when the number of possible modifications is large, and the monolithic simulation is inefficient from the computational cost perspective. Through hierarchical approach one could explore the space of possible modifications in a much more cost efficient manner.

6   Conclusions and Future Work

We have shown how a hierarchical approach can be used to abstract from a fine grain agent based simulation up to a mean field approximation that can effectively remove the dependency on the number of agents of the computational cost.

Our approach moves the difficulty of the solution to crowd movement in evacuation context from the high scale computational cost to a high number of states in a system of parallel DTMCs. This tradeoff allows for faster calculation of the solution, and offers a deterministic solution through mean field approximation.

As for the future development, we are interested in a formalisation of a parallel Markov chain models with variable number of abstractions, like the physical, social and regional aspect in this work, and its implementation in already available tools that lack this feature. Another interesting aspect would be the formalisation of conditional transitions for single DTMCs that would allow for a more rich behaviour of the model while maintaining the mathematical validity for the application of already existing theoretical approaches.

References

1. Massink, M., Latella, D., Bracciali, A., Harrison, M.D.: A scalable fluid flow process algebraic approach to emergency egress analysis. In: 8th IEEE International Conference on Software Engineering and Formal Methods (SEFM), pp. 169–180. IEEE (2010)
2. von Sivers, I., Templeton, A., Köster, G., Drury, J., Philippides, A.: Humans do not always act selfishly: social identity and helping in emergency evacuation simulation. Transp. Res. Procedia **2**, 585–593 (2014)
3. Helbing, D., Farkas, I., Vicsek, T.: Simulating dynamical features of escape panic. Nature **407**(6803), 487–490 (2000)

4. Gorrini, A., Bandini, S., Vizzari, G.: Empirical investigation on pedestrian crowd dynamics and grouping. In: Chraibi, M., Boltes, M., Schadschneider, A., Seyfried, A. (eds.) Traffic and Granular Flow '13, pp. 83–91. Springer, Heidelberg (2015)

5. Khan, S.D., Vizzari, G., Bandini, S.: Identifying sources and sinks and detecting dominant motion patterns in crowds. Transp. Res. Procedia 2, 195–200 (2014)

6. Helbing, D., Johansson, A., Al-Abideen, H.Z.: Dynamics of crowd disasters: an empirical study. Phys. Rev. E 75(4), 046109 (2007)

7. Narain, R., Golas, A., Curtis, S., Lin, M.C.: Aggregate dynamics for dense crowd simulation. ACM Trans. Graph. (TOG) 28, 122 (2009). ACM

8. Dogbe, C.: On the modelling of crowd dynamics by generalized kinetic models. J. Math. Anal. Appl. 387(2), 512–532 (2012)

9. Templeton, A., Drury, J., Philippides, A.: From mindless masses to small groups: conceptualizing collective behavior in crowd modeling. Rev. Gen. Psychol. 19(3), 215–229 (2015)

10. Wijermans, N., Jorna, R., Jager, W., van Vliet, T., Adang, O.: Cross: modelling crowd behaviour with social-cognitive agents. J. Artif. Soc. Soc. Simul. 16(4), 1 (2013)

11. Bagnoli, F., Rechtman, R.: Bifurcations in models of a society of reasonable contrarians and conformists. Phys. Rev. E 92(4), 042913 (2015)

12. Le Boudec, J-Y., McDonald, D., Mundinger, J.: A generic mean field convergence result for systems of interacting objects. In: Fourth International Conference on the Quantitative Evaluation of Systems, QEST 2007, pp. 3–18. IEEE (2007)

13. Latella, D., Loreti, M., Massink, M.: On-the-fly PCTL fast mean-field approximated model-checking for self-organising coordination. Sci. Comput. Program. 110, 23–50 (2015)

14. Moussaïd, M., Helbing, D., Theraulaz, G.: How simple rules determine pedestrian behavior and crowd disasters. Proc. Nat. Acad. Sci. 108(17), 6884–6888 (2011)

Modeling and Simulation Tools

Simulating Hybrid Systems Within SIMTHESys Multi-formalism Models

Enrico Barbierato[1]([✉]), Marco Gribaudo[1], and Mauro Iacono[2]

[1] Dip. di Elettronica, Informazione e Bioingegneria, Politecnico di Milano,
via Ponzio 34/5, 20133 Milano, Italy
{enrico.barbierato,marco.gribaudo}@polimi.it
[2] Dip. di Matematica e Fisica, Seconda Università degli Studi di Napoli,
viale Lincoln 5, 81100 Caserta, Italy
mauro.iacono@unina2.it

Abstract. As many real world systems evolve according to phenomena characterized by a continuous time dependency, literature studied several approaches to correctly capture all their aspects. Since their analysis is not trivial, different high level approaches have been proposed, such as classical pure mathematical analysis or tool-oriented frameworks like Fluid Stochastic Petri Nets. Each approach has its specific purposes and naturally addresses some application field. This paper instead focuses on the simulation of models written in a custom Hybrid Systems (HS) formalism. The key aspect of this work is focused on the use within a framework called SIMTHESys of a function describing how the fluid variables evolve, providing more efficient simulation with respect to traditional approaches.

Keywords: Hybrid systems · Simulation · SIMTHESys

1 Introduction

Real world systems frequently exhibit dynamics emerging from a combination of events occurring in well-defined instants of time and phenomena evolving continuously during time. This hybrid nature of non-trivial systems is typical of computer controlled physical processes, e.g. in industrial manufacturing or chemical plants. Modeling and analysis of these systems should preserve the actual relations between the continuous and discrete time behavior, as a separate approach for the two aspects may introduce artificial effects or hide existing ones. Literature dubbed such systems Hybrid Systems (HS). Different approaches have been proposed in the literature to cope with specific applications. In this paper, rather that providing another formalism, we leverage the SIMTHESys approach [14] to provide a general foundation for the development of custom modeling formalisms supporting user defined HS description languages, and propose the theoretical issues that are behind its new HS solving engine. By means of similar integrations, SIMTHESys enables the implementation of very different, rich

D. Fiems et al. (Eds.): EPEW 2016, LNCS 9951, pp. 189–203, 2016.
DOI: 10.1007/978-3-319-46433-6_13

high level modeling formalisms: some significant examples are given by a version of stochastic automata for condition testing [2], a Domain Specific Language for multithreaded, multiprocessor systems [3] and general approaches allowing product form solutions [6] and exceptions support [5]. Several methods (both analytical and simulative) have been proposed previously to exploit solutions of hybrid models. In this work we follow the representation used by the SIMTHESys Hybrid Formalism Family (HFF), introduced in [4] and present the SIMTHESys Hybrid Formalism Solving Engine (HFSE) simulation component. With respect to similar approaches, where the evolution of a fluid model is described through ODEs, this method exploits a function describing in a deterministic way the behavior of the fluid model. The advantage of this approach consists of a faster resolution of the model, though it requires a modification of the existing techniques used by SIMTHESys.

The paper is organized as follows: in Sect. 2 background and related works are presented; Sect. 3 describes the modeling approach; Sect. 4 details the simulation technique; Sect. 5 describes some applications, followed by conclusions.

2   Background and Related Works

The interest for HS stems from architectural designs involving cooperating objects, reflecting both discrete and continues dynamics. A stimulating introduction to HS is provided by [1], which illustrates the foundations of the on-going research and the relevant tools developed so far. Another interesting resource can be found in [8], which takes in account a survey of languages and tools managing hybrid systems. The authors stress the importance of the assumptions related to environment and the difficulty in sharing information among tools, considering some crucial factors for their evaluation: i) the expressive power, ii) the mathematical theory representing the theoretical basis of the language and iii) the underlying semantics, the latter being probably the most relevant to drive the tool development. Among the different tools, some of them offer interesting perspectives, notably UPPAAL [16], HyTech [13], Ptolemy [18] and Prism [15].

Issues related to embedded systems (mainly safety-critical applications) motivated the study of Hybrid Automata [12]; significant cases of their application can be found, for example, in [7] and [20].

SIMTHESys (Structured Infrastructure for Multiformalism modeling and Testing of Heterogeneous formalisms and Extensions for SYStems) provides a set of solution engines able to compute performance indexes of a multiformalism model, either by numerical algorithms or by simulation. Formalisms can be grouped in families (depending on the techniques requested to study the model). Their modeling primitives are described by Formalism Description Language (FDL, based on XML) files, including one or more *Elements* which specifies in turn a set of *Properties* and *Behaviors*: the former describes the performance indices computed by the engines, the latter the semantic of the *Element*. Two or more elements can share one or more behaviors by using a set of *Interfaces*. *Models* are described by MDL (Model Description Language) according to the

FDL file. SIMTHESys users are provided a set solution engines for different formalism families (i.e. exponential events, exponential and immediate events based and labeled exponential events formalisms), including both discrete event simulation and state space generation.

3 Modeling Approach

The language used to describe the HS is a subset of the Piecewise Deterministic Markov Processes introduced in [9], which has been addressed in [4] as Hybrid System Modeling Language (HSML). In particular, an HS is characterized by a discrete and finite set of modes $\mathcal{M} = \{m_1, \ldots, m_M\}$. Every mode m_i has a finite number d_i of continuous *variables* $x_{i,j}$ (with $1 \leq j \leq d_i$) that are defined on a compact subset of $\mathbb{R}$ contained between by a lower boundary $l_{i,j}$ and an upper boundary $u_{i,j}$. Each mode m_i is thus characterized by a *continuous domain* $D_i = \times_{j=1}^{d_i} [l_{i,j}, u_{i,j}]$, where $\times$ represents the cartesian product of the corresponding sets. States are then defined by a tuple $\sigma = (m_i, \mathbf{x}_i)$, where $\mathbf{x}_i = (x_{i,1}, \ldots, x_{i,d_i})$ denotes the values of all the continuous variables of the corresponding mode. The model evolves through a state space $\mathcal{S} = \bigcup_{i=1}^{M} (\{m_i\} \times D_i)$ with $\sigma \in \mathcal{S}$.

A tuple $(\mathcal{S}, \Phi, E, \Lambda, \Psi, \sigma_0)$ provides the definition of a HSML model evolving on the state space $\mathcal{S}$. The evolution of the continuous components of the states is encoded in functions $\Phi = \{\phi_1, \ldots, \phi_M\}$. A different function $\phi_i : D_i \times \mathbb{R} \rightarrow D_i$ is given for each mode $m_i \in \mathcal{M}$. If we consider a time interval $[t_a, t_b]$ where the system remains in the same mode m_i, we can use function ϕ_i to describe the temporal evolution of the continuous variables. We denote with $\sigma(t) = (m_i, \mathbf{x}_i(t))$ the state at time t, with $t_a \leq t \leq t_b$. Note that since by hypothesis the system remains in mode m_i in the considered time interval, only the continuous component of the state depends on t. Then, $\sigma(t_c) = (m_i, \mathbf{x}_i(t_c)) \implies \sigma(t_d) = (m_i, \phi_i(\mathbf{x}_i(t_c), t_d - t_c))$, $\forall t_a \leq t_c \leq t_d \leq t_b$. By definition we have that $\phi_i(\mathbf{x}_i(t), 0) = \mathbf{x}_i(t)$.

Beside the continuous evolution of the model defined by functions belonging to Φ, the state $\sigma(t)$ can also be modified by the occurrence of one of N possible *events* $E = \{e_1, \ldots, e_N\}$. A function $\Lambda : S \times E \rightarrow \mathbb{R}_0^+$ defines the state dependent rate at which events occur. In detail, $Pr\{e_k$ occurs in state $\sigma(t)$ during $\Delta t\} = \Lambda(\sigma(t), e_k) \cdot \Delta t + o(\Delta t)$. The effect of the event is described by a function $\Psi : S \times E \times S \rightarrow [0, 1]$. Specifically, $\Psi(\sigma(t^+)|e_k, \sigma(t^-))$ defines the probability that the state $\sigma(t^-)$ becomes $\sigma(t^+)$ due to the occurrence of event e_k. The change of state can affect both the mode and the continuous variables in a probabilistic way: for this reason function Ψ is defined such that $\Psi(\sigma(t^+)|e_k, \sigma(t^-)) = Pr\{\sigma(t^+) = (m_j, \mathbf{x}')$ with $x'_{j,1} \leq x_{j,1}, \ldots, x'_{j,d_j} \leq x_{j,d_j}|e_k, \sigma(t^-)\}$.

As a short-hand notation, we write $\Psi(e_k, \sigma(t)) = DIST(\sigma(t))$, using $DIST(\sigma(t))$ to identify a probability distribution that can depend on the current state $\sigma(t)$. For instance, $\Psi(e_k, \sigma(t) = (m_i, \mathbf{x}_i)) = DET(m_l, \mathbf{x}_i)$ denotes the effect of an event e_k that changes deterministically (DET) the mode from m_i to m_l while leaving the continuous variables unchanged. Finally, $\sigma_0 \in \mathcal{S}$ represents the initial state of the model.

3.1 HSML vs. Fluid Models

Stochastic Fluid Models [19] have been widely used in the literature to describe hybrid systems. The main difference between fluid models and HSML lies in the way in which the evolution of the continuous variables is described. In fluid models, each discrete state m_i (that is the equivalent of an HSML mode), is characterized by a set of rates $r_{i,j}(\sigma(t))$. Continuous variables x_j evolve when the model is in state $\sigma(t)$ according to the following system of ordinary differential equations:

$$\frac{dx_j}{dt} = r_{i,j}(\sigma(t)), \quad \forall 1 \leq j \leq d_i. \tag{1}$$

The value of the continuous variables at the time in which the system enters a new state provides the initial value for the ODEs: the value assumed by the continuous variables as function of time during a trace is computed by integrating Eq. 1. In HSML instead functions Φ define the evolution of the model as function of time. In practice, $\phi_i(\mathbf{x}, t)$ contains the solution of an equation like (1) at time t, with an initial value for the continuous variable equal to $\mathbf{x}$. Although this difference might seems to be just notational, it has an impact on the solution techniques and on the model specification that can be given: in many applications, function ϕ_i can be easier to define than the rate $r_{i,j}$ (some examples will be discussed in Sect. 5). As will be detailed in Sect. 4, the use of functions Φ can also simplify the simulation of cases that would be much harder to study when expressed with conventional fluid models.

4 Discrete Event Simulation of HS

The proposed HSML models can be analyzed both numerically and using discrete event simulation. However, due to the hybrid nature including both discrete and continuous quantities, both types of techniques are very complex to implement. In this work we will focus only on discrete event simulation.

Since the event rate Λ depends on the continuous evolution of the state, the underlying stochastic process is non-homogenous. Two techniques that are well-suited for this type of systems are *thinning* and *time-scale transformation* [17]: both were compared in [10] to simulate Fluid Stochastic Petri Nets. In this work we mainly focus on time-scale transformation.

4.1 Time-Scale Transformation

Time-scale transformation is based on the following result. Given an event $e \in E$, let us call $H_e(t)$ the integral of its firing rate:

$$H_e(t) = \int_0^t \Lambda(\sigma(\tau), e) d\tau \tag{2}$$

The time-scale transformation technique exploits the fact that $H_e(t)$ is distributed as an exponential distribution of rate 1.

In HSML, since change of modes can only occur when events are fired, supposing i)that between two events the system remains in mode m_i, and ii)that at the time in which the previous event occurred, the value of the continuous variables was $\mathbf{x}_i$ then, thanks to the description of the continuous evolution in HSML models based on functions Φ, Eq. 2 can be rewritten as:

$$H_e(t|m_i, \mathbf{x}_i) = \int_0^t \Lambda((m_i, \phi_i(\mathbf{x}_i, \tau)), e)d\tau \qquad (3)$$

The time Δt_e at which event e will occur after entering into state $\sigma_i = (m_i, \mathbf{x}_i)$, can then be computed by first drawing a sample $\xi \sim EXP(1)$ from an exponential distribution of rate 1, and then inverting Eq. 3:

$$\Delta t_e = H_e^{-1}(\xi_e|m_i, \mathbf{x}_i) \qquad (4)$$

Although in general Eq. 4 should be inverted numerically, there are many practical cases in which it could be done in closed form starting from the definitions of Φ and Λ: some of them will be presented in Sect. 5. Moreover, even if Eq. 4 cannot be inverted directly, but closed form expressions exist for $\Lambda((m_i, \phi_i(\mathbf{x}_i, \tau)), e)$ and Eq. 3, Δt_e can still be computed quite efficiently using the Newton-Raphson method since $H_e(t)$ is monotonically increasing and its derivative is known.

Algorithm 1. SimulateHSML($[S, \Phi, E, \Lambda, \Psi, \sigma_0], T_{\max}$)

1: $(m, \mathbf{x}) = \sigma_0; t = 0$
2: **while** $t < T_{\max}$ **do**
3:    **for all** $e \in E$ **do**
4:        $\xi_e \sim EXP(1)$
5:        $\Delta t_e = H_e^{-1}(\xi_e|m, \mathbf{x})$
6:    **end for**
7:    $e = \arg\min_{e \in E} \Delta T_e$
8:    $t = t + \Delta T_e$
9:    $\mathbf{x} = \phi_m(\mathbf{x}, \Delta T_e)$
10:   $(m, \mathbf{x}) \sim \Psi(e, m, \mathbf{x})$
11: **end while**

To summarize, Algorithm 1 performs one simulation run of an HSML model from time $t = 0$ up to time $t = T_{\max}$ with time-scale transformation. The initial time and states are set at line 1. The iteration until the maximum time $T_{\max}$ is reached is performed at line 2. The firing times of the events are computed in the iteration starting at line 3. Then the event e with the minimum firing time is chosen to fire at line 7. Starting from line 8 the algorithms updates the simulation time and the values of the continuous variables in the current mode. The new state is computed at line 10 by sampling a new value for both the modes and the continuous variables in the new state from the distribution defined by function $\Psi(e, m, \mathbf{x})$.

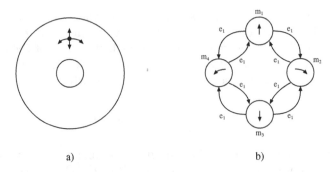

Fig. 1. A moving agent in a circular pattern: (a) the system; (b) the corresponding HSML model.

4.2 Performance Indices

When studying HS, generally the interest focuses on performance measures involving the distribution of the continuous variables though providing confidence intervals of continuous distributions estimated with Monte-Carlo techniques is a research topic on its own. Although some techniques exist, such for example considering the probability of belonging to a small interval as a Bernoulli variable, they are very complex and outside the scope of this paper.

5 Applications

To better clarify the simulation approach, this section presents three examples: the first two are artificial models, specifically created to emphasize some peculiarities of the solution technique; the last one is instead an application to the analysis of urban transport.

5.1 Moving Agent with a Circular Pattern

First, a model of an agent moving through circular patterns is considered, as shown in Fig. 1a. In particular, the agent is characterized by four modes $m_1, \ldots, m_4$ representing respectively the motion toward the border, the clockwise turn, the movement versus the centre, and the counter-clockwise turn. Each mode is characterized by two continuous variables x_1 and x_2 that represents respectively the horizontal and vertical coordinates of the position of the agent on the plane. The corresponding state space is thus defined as $S = \{m_1, m_2, m_3, m_4\} \times [-1, 1] \times [-1, 1]$.

The agent moves at a constant tangential speed of μ units per second, either linearly or tangential to the circle. The agent cannot move further than 1 unit from the centre of the circle, and not closer than a minimum radius $r_{\min}$: if one of the boundary is reached, the agent stops moving and waits in the position for a mode change. Let us call $\mathbf{x} = (x_1, x_2)$. By indicating with $\rho(\mathbf{x})$ and $\alpha(\mathbf{x})$

respectively the distance from the centre and the counter-clockwise angle from the positive horizontal axis, the functions $\Phi = \{\phi_1, \ldots \phi_4\}$ can be defined as:

$$\phi_1(\mathbf{x}, \Delta t) = \begin{pmatrix} \min(1, (\rho(\mathbf{x}) + \mu\Delta t)) \cdot \cos(\alpha(\mathbf{x})) \\ \min(1, (\rho(\mathbf{x}) + \mu\Delta t)) \cdot \sin(\alpha(\mathbf{x})) \end{pmatrix}^T \tag{5}$$

$$\phi_2(\mathbf{x}, \Delta t) = \begin{pmatrix} \rho(\mathbf{x}) \cdot \cos(\alpha(\mathbf{x}) - \frac{\mu\Delta t}{\rho(\mathbf{x})}) \\ \rho(\mathbf{x}) \cdot \sin(\alpha(\mathbf{x}) - \frac{\mu\Delta t}{\rho(\mathbf{x})}) \end{pmatrix}^T \tag{6}$$

$$\phi_3(\mathbf{x}, \Delta t) = \begin{pmatrix} \max(r_{\min}, (\rho(\mathbf{x}) - \mu\Delta t)) \cdot \cos(\alpha(\mathbf{x})) \\ \max(r_{\min}, (\rho(\mathbf{x}) - \mu\Delta t)) \cdot \sin(\alpha(\mathbf{x})) \end{pmatrix}^T \tag{7}$$

$$\phi_4(\mathbf{x}, \Delta t) = \begin{pmatrix} \rho(\mathbf{x}) \cdot \cos(\alpha(\mathbf{x}) + \frac{\mu\Delta t}{\rho(\mathbf{x})}) \\ \rho(\mathbf{x}) \cdot \sin(\alpha(\mathbf{x}) + \frac{\mu\Delta t}{\rho(\mathbf{x})}) \end{pmatrix}^T \tag{8}$$

Mode change is governed by a single event e_1, that can occur in every mode at constant Poisson rate λ, which corresponds to:

$$E = \{e_1\}, \quad \Lambda(\sigma, e_1) = \lambda \quad \forall \sigma \in \mathcal{S} \tag{9}$$

When the event occurs, the mode switch from radial to angular motion, randomly choosing (with half and half probability) the direction. This is expressed by defining function Ψ as follows:

$$\Psi(e_1, m, \mathbf{x}) = \begin{cases} (m_2, \mathbf{x}) \text{ with prob. } 0.5 \\ (m_4, \mathbf{x}) \text{ with prob. } 0.5 \end{cases} \quad \text{if } m \in \{m_1, m_3\} \tag{10}$$

$$\Psi(e_1, m, \mathbf{x}) = \begin{cases} (m_1, \mathbf{x}) \text{ with prob. } 0.5 \\ (m_3, \mathbf{x}) \text{ with prob. } 0.5 \end{cases} \quad \text{if } m \in \{m_2, m_4\} \tag{11}$$

Finally, the system starts rotating counter-clockwise, at half of the maximum radius on the positive vertical axis, that is: $\sigma_0 = (m_4, 0, 0.5)$. The HSML is summarized in Fig. 1b, and the parameters used in the experiments are the following: $\mu = 0.1, \lambda = 2, r_{\min} = 0.1$.

Although the system is very simple, circular motion is very hard to describe using conventional fluid models. In particular, the circular motion should have been defined using a fluid dependent rate, and it would have been computed by solving the following ODE:

$$\frac{d\mathbf{x}}{dt} = \begin{pmatrix} \mu \cdot \cos(\alpha(\mathbf{x}) + \pi/2) \\ \mu \cdot \sin(\alpha(\mathbf{x}) + \pi/2) \end{pmatrix}^T \tag{12}$$

Equation 12 is unfortunately very hard to solve numerically: simple techniques such as Euler produce results that tend to diverge even with very small discretization steps, and more advanced techniques such as adaptive step-size Runge-Kutta are needed as shown in Fig. 2a. The definition of motion using function Φ in HSML avoids this problem by considering directly the solution. Figure 2b shows three possible execution traces of the model, considering different seeds. The fluid density over the space for different time instants are shown

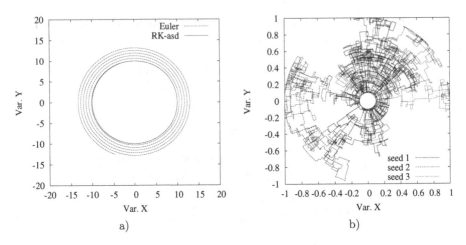

Fig. 2. A moving agent in a circular pattern: (a) integration of a circular motion; (b) three evolution traces.

in Fig. 3. As it can be seen in Fig. 3a, at the beginning, the system is influenced by the probability mass deriving from the initial state. As time increases, the probability starts to distribute in a more uniform way, even if the initial state creates asymmetries that are destroyed only after very long time ($T = 1000$). It is also interesting to note that the upper and lower boundary to the radius creates two borders where probability mass accumulates. Depending on the length of the simulation interval, the technique was able to consider 10^8 traces in around one minute, and down-scaled to $5 \cdot 10^6$ traces in around ten minutes for $T = 1000$ on a 2011 MacBook Air, with 4GB of RAM and using a single core of an Intel i5 CPU.

5.2   Leaky Bucket with Batch Arrivals

The second example aims at showing other features of HSML models. We consider a leaky bucket model of a streaming application, where data arrives in batches, as shown in Fig. 4a. The system has three modes. Mode m_2 represents the normal operation of the system: it is characterized by a continuous variable $x \in [0, 1]$ representing the level of the bucket. The other two modes m_1 and m_3 represent respectively the full and the empty bucket. The state space can thus be defined as $\mathcal{S} = \{m_1, m_3\} \cup (\{m_2\} \times [0, 1])$.

Continuous variables are only present in m_2, and they uniformly decrease at unitary rate, leading to the following definition of $\Phi = \{\phi_2\}$:

$$\phi_2(x, \Delta t) = \max(0, x - \Delta t) \tag{13}$$

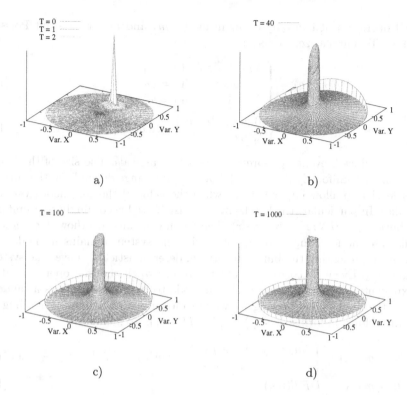

Fig. 3. Position probability of a moving agent in a circular pattern: (a) T = 1,2,3; (b) T = 40; (c) T = 100; (d) T = 1000.

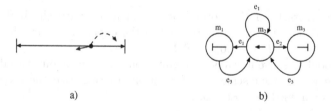

Fig. 4. A leaky bucket with batch arrivals: (a) the system; (b) the corresponding HSML model.

The model is characterized by three events $E = \{e_1, e_2, e_3\}$. Event e_1 represents the batch arrivals: they occur only in mode m_2, they are allowed to arrive at a higher frequency when the buffer is empty, and they are slowed down when the system is full. In particular, with a buffer level x, event e_1 occurs at speed $\lambda_J \cdot (1 - x)^\alpha$. Event e_2 fires deterministically when the system reaches the empty buffer in mode m_2: as in many other fluid model formalisms, we can use Dirac's delta $\delta(x)$ to denote events that deterministically occur when a specific value of the continuous variables is reached. Finally, event e_3 represents the return to normal operation: it can only occur in modes m_1 and m_3 at constant Poisson speed λ_r. To summarize, we have:

$$\Lambda(m_2, x, e) = \begin{cases} \lambda_J \cdot (1-x)^\alpha & \text{if } e = e_1 \\ \delta(x) & \text{if } e = e_2 \\ 0 & \text{if } e = e_3 \end{cases} \tag{14}$$

$$\Lambda(m, e) = \begin{cases} 0 & \text{if } e \in \{e_1, e_2\} \\ \lambda_r & \text{if } e = e_3 \end{cases} \quad \text{if } m \in \{m_1, m_3\} \tag{15}$$

As introduced, event e_1 represents the batch arrivals. The size of the batch is random and uniformly distributed, however, the range depends on the current buffer level x to allow larger batches when the value of the continuous variable is smaller. In particular, the batch size y is distributed accordingly to a uniform distribution $y \sim UNIF(0, \beta - \gamma \cdot x)$. The batch can cause overflow if $x + y > 1$: in this case mode changes to m_1, otherwise the system remains in mode m_1. Event e_2, representing the buffer underflow, deterministically moves the system to mode m_3. Event e_3 corresponds to the return to the normal operation after an exponential delay changing back the mode to m_2. It also injects a random level in the fluid buffer, setting its value to a sample distributed according to the sum of four uniforms: $x \sim \frac{1}{4}\sum_{i=1}^{4} UNIF(0, 1)$. To summarize:

$$\Psi(e_1, m_2, \mathbf{x}) = \begin{cases} (m_2, x+y) & \text{if } x + y < 1 \\ m_1 & \text{if } x + y \geq 1 \end{cases} \quad \text{with } y \sim UNIF(0, \beta - \gamma \cdot x) \tag{16}$$

$$\Psi(e_2, m_2, \mathbf{x}) = DET(m_3) \tag{17}$$

$$\Psi(e_3, m) = \left(DET(m_2), \frac{1}{4}\sum_{i=1}^{4} UNIF(0,1) \right) \quad \text{if } m \in \{m_1, m_3\} \tag{18}$$

The system starts from the normal operation mode with the buffer half-full $\sigma_0 = (m_2, 0.5)$, and the parameters used in the experiments are the following: $\lambda_r = 1, \lambda_J = 10, \alpha = 2, \beta = 0.6, \gamma = 0.2$.

This model has event rates depending on the continuous variables, so it requires the use of time-scale transformation. In this case the integral firing rate can be expressed in closed form:

$$H_{e_1}(t|m_2, x) = \begin{cases} I(x|x) + \lambda_J \cdot (t - x) & \text{if } t \geq x \\ I(t|x) & \text{if } t < x \end{cases} \tag{19}$$

where

$$I(t|x) = \int_0^t \lambda_J \cdot (1 - x + \tau)^\alpha \, d\tau = \frac{\lambda_J}{\alpha + 1} \left[(1 - x + \tau)^{\alpha+1} - (1 - x)^{\alpha+1} \right] \tag{20}$$

Equation 19 can also be inverted in closed form:

$$H_{e_1}^{-1}(\xi|m_2, x) = \begin{cases} x - 1 + \sqrt[\alpha+1]{\xi \frac{\alpha+1}{\lambda_J} + (1-x)^{\alpha+1}} & \text{if } \xi \leq \frac{\lambda_J}{\alpha+1}\left[1 - (1-x)^{\alpha+1}\right] \\ x + \frac{\xi}{\lambda_J} - \frac{1-(1-x)^{\alpha+1}}{\alpha+1} & \text{if } \xi > \frac{\lambda_J}{\alpha+1}\left[1 - (1-x)^{\alpha+1}\right] \end{cases} \tag{21}$$

The computation of the firing time of event e_2 can then be done by generating an exponentially distributed sample $\xi \sim EXP(1)$ and then applying Eq. 21. By exploiting this feature, 10^8 traces up to $T = 2$ could be evaluated in less than four minutes on the same PC described in Sect. 5.1. Figure 5 presents three different traces, showing both the mode and the value for the continuous variable (for mode m_2).

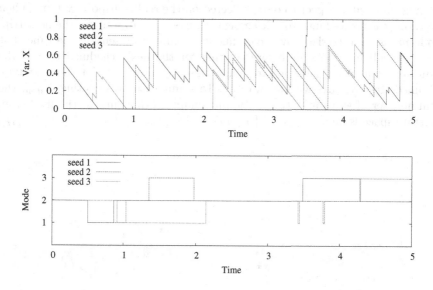

Fig. 5. A leaky bucket with batch arrivals: three evolution traces.

Figure 6 shows the evolution of the distribution of the continuous variable (Fig. 6a) and of the probability mass (Fig. 6b). As it can be seen, the probability mass is accumulated in modes m_1 and m_3 since they are not characterized by any continuous variable. However, there is also an evolving probability mass in state m_2 up to time $T = 0.5$ to consider the influence of the initial state of the model.

5.3 An Application to the Urban Transport System

The third case study presents a multi-formalism model of a tram route in an urban environment. The route of a tram is modeled as an HSML building block in a more complex application composing it with a Fluid Stochastic Petri Net with Flush-out Arcs (FSPNs) [11]. The considered tram travels on a circular route: a continuous variable is used to describe the position on the route, assuming integer values at the stops, and real values between them (see Fig. 7a). The purpose of this model is not providing a full-fledge application to analyze urban transport system (for which there is a broad literature providing very interesting solutions), but to give an idea of which type of applications can be supported by describing HS with multi-formalism models. The corresponding HSML, shown in Fig. 7b, is defined by four modes $m_1, \ldots, m_4$. The main idea is to collect three timetables for the tram from real measurements: the one with the fastest running time, the one with the average running time, and the one with the slowest running time. Each timetable is followed respectively in modes m_1, m_2 and m_3, and the HSML model alternates randomly between the three. The jump rates between the modes are chosen to keep the average and the variance of the running time with respect to data collected on the real transport system. When the tram reaches its final stop, it changes to mode m_4 to wait for the next trip. Mode m_4 is characterized by a continuous variable representing the time that the tram will have to wait before starting for the next scheduled trip. If the tram arrives at the final stop later than the scheduled time for the next trip, it immediately restarts without waiting. To summarize, by denoting N_{stops} the total number of stops and T_{wmax} the maximum waiting time at the final stop, the state space is defined as $\mathcal{S} = \{m_1, m_2, m_3\} \times [0, N_{stops}] \cup \{m_4\} \times [0, T_{wmax}]$.

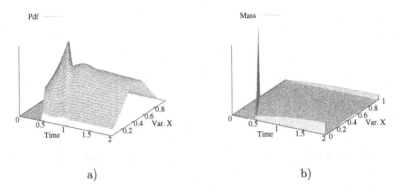

Fig. 6. A leaky bucket with batch arrivals: (a) probability density of the value of variable X; (b) probability mass of X.

Function ϕ_i with $i \in \{1, 2, 3\}$, reflects the movement of the tram according to the time table. Let us call $\eta_i(t)$ the function that for the i-th mode defines the value of the continuous variable that identifies the position of the bus at time t, then ϕ_i is defined as:

$$\phi_i(x, \Delta t) = \begin{cases} \min\left(N_{stops}, \eta_i(\eta_i^{-1}(x) + \Delta t)\right) & i \in \{1, 2, 3\} \\ \max(0, x - \Delta t) & i = 4 \end{cases} \tag{22}$$

The evolution of the model is governed by three events $E = \{e_1, e_2, e_3\}$. Event e_1 represents the change of timetable in modes m_1, m_2 or m_3. Event e_2 is fired when the tram reaches the final stop: it either brings the system to mode m_4 if the tram is on time, or it remains in the other modes if it arrives late with respect to the schedule. Finally, event e_3 represents the start of the next trip when the system is waiting in mode m_4. Because of space constraint, Λ and Ψ are not defined.

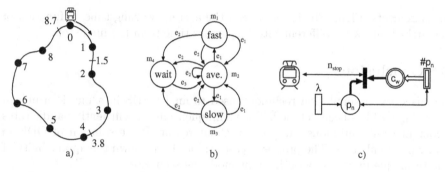

Fig. 7. An urban transport: (a) the system; (b) the corresponding HSML model; (c) multi-formalism application

Figure 7c shows a multi-formalism model exploiting FSPNs to compute the average waiting time at a given stop. A discrete transition with rate λ pushes tokens representing travellers in discrete place p_n. A state-dependent fluid transition firing at a rate identical to the number of tokens in p_n pumps fluid in place c_w. In this way, place c_w accounts for the total waiting time (i.e. the integral of the number of waiting costumers over time). Whenever the tram reaches the stop number n_{stop}, it enables an immediate transition flushing out both place p_n and c_w. In this way, the average waiting time can be computed as the ratio of the fluid flushed out from place c_w, with the number of tokens removed from place p_n.

Timetables with real data were considered, specifically the one corresponding to tram 16 from the city of Torino (Italy), using timetables available on the web. Figure 8a shows a trace of the model, and compares it with the fastest, the slowest, the average timetables of the tram, plus one trace collected from the real system. As we can see the synthetic trace is compatible with the real

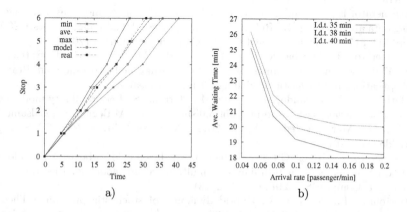

Fig. 8. An application of multi-formalism modeling to urban transport: (a) comparison of real and synthetic traces; (b) average travellers waiting time.

measurements. Figure 8b shows instead the average waiting time as function of the arrival rate λ for different inter-departure time from the final stop.

6   Conclusions

In this work, a simulation technique used to analyze HS in a multi-formalism modeling tool has been detailed. The technique is able to efficiently handle rates depending on continuous variables, without requiring the solution of ODE in many practical cases. The proposed case studies have shown the applicability of the technique in a variety of different modeling scenarios.

Future work will focus on numerical techniques to solve the Partial Differential Equations (PDE) that describes the same set of HS in an analytical way, and on the determination of the accuracy and the convergence to a given confidence level for simuliation approaches. Practical case study will include fluid models of Big-Data applications, transportation systems and energy consumption in datacentres.

Acknowledgements. The results of this work have been partially funded by EUBra-BIGSEA (grant agreement no. 690116), a Research and Innovation Action (RIA) funded by the European Commission under the Cooperation Programme, Horizon 2020 and the Ministrio de Cincia, Tecnologia e Inovao (MCTI), RNP/Brazil (grant GA0000000650/04).

References

1. Antsaklis, P.J., Koutsoukos, X.D.: Hybrid Systems: Review and Recent Progress, pp. 273–298. Wiley, Hoboken (2005)
2. Barbierato, E., Gribaudo, M., Iacono, M.: Exploiting multiformalism models for testing and performance evaluation in SIMTHESys. In: Proceedings of 5th International ICST Conference on Performance Evaluation Methodologies and Tools - VALUETOOLS 2011 (2011)
3. Barbierato, E., Gribaudo, M., Iacono, M.: A performance modeling language for big data architectures. In: Rekdalsbakken, W., Bye, R.T., Zhang, H. (eds.) ECMS, pp. 511–517. European Council for Modeling and Simulation (2013)
4. Barbierato, E., Gribaudo, M., Iacono, M.: Modeling hybrid systems in SIMTHESys. In: Proceedings of the 8th International Workshop on Practical Applications of Stochastic Modelling (2016, to appear)
5. Barbierato, E., Gribaudo, M., Iacono, M., Marrone, S.: Performability modeling of exceptions-aware systems in multiformalism tools. In: Al-Begain, K., Balsamo, S., Fiems, D., Marin, A. (eds.) ASMTA 2011. LNCS, vol. 6751, pp. 257–272. Springer, Heidelberg (2011)
6. Barbierato, E., Rossi, G.L.D., Gribaudo, M., Iacono, M., Marin, A.: Exploiting product forms solution techniques in multiformalism modeling. Electron. Notes Theoret. Comput. Sci. **296**, 61–77 (2013)
7. Bortolussi, L., Policriti, A.: Hybrid dynamics of stochastic programs. Theoret. Comput. Sci. **411**(20), 2052–2077 (2010). http://www.sciencedirect.com/science/article/pii/S0304397510001039, hybrid Automata and Oscillatory Behaviour in Biological Systems

8. Carloni, L.P., Passerone, R., Pinto, A., Sangiovanni-Vincentelli, A.L.: Languages and tools for hybrid systems design. Found. Trends Electron. Des. Autom. **1**(1/2), 1–193 (2006). http://dx.doi.org/10.1561/1000000001

9. Davis, M.: Markov Models & Optimization. Chapman & Hall/CRC Monographs on Statistics & Applied Probability. Taylor & Francis, Abingdon (1993)

10. Gribaudo, M., Sereno, M.: Simulation of fluid stochastic petri nets. In: 8th International Symposium on Modeling, Analysis and Simulation of Computer and Telecommunication Systems, Proceedings, pp. 231–239 (2000)

11. Gribaudo, M., Sereno, M., Horváth, A., Bobbio, A.: Fluid stochastic petri nets augmented with flush-out arcs: modelling and analysis. Discrete Event Dyn. Syst. **11**(1–2), 97–117 (2001). http://dx.doi.org/10.1023/A:1008339216603

12. Henzinger, T.A.: The theory of hybrid automata. In: Proceedings of the 11th Annual IEEE Symposium on Logic in Computer Science, LICS 1996, p. 278. IEEE Computer Society, Washington, DC (1996)

13. Henzinger, T.A., Ho, P.H., Wong-toi, H.: Hytech: a model checker for hybrid systems. Softw. Tools Technol. Transf. **1**, 460–463 (1997)

14. Iacono, M., Gribaudo, M.: Element based semantics in multi formalism performance models. In: MASCOTS, pp. 413–416. IEEE (2010)

15. Kwiatkowska, M., Norman, G., Parker, D.: Probabilistic symbolic model checking with PRISM: a hybrid approach. Int. J. Softw. Tools Technol. Transf. **6**(2), 128–142 (2004)

16. Larsen, K.G., Pettersson, P., Yi, W.: Diagnostic model-checking for real-time systems. In: Alur, R., Henzinger, T.A., Sontag, E.D. (eds.) Proceedings of Workshop on Verification and Control of Hybrid Systems III. LNCS, vol. 1066, pp. 575–586. Springer, Heidelberg (1995)

17. Lewis, P.A.W., Shedler, G.S.: Simulation of nonhomogeneous poisson processes by thinning. Naval Res. Logistics Q. **26**(3), 403–413 (1979). doi:10.1002/nav. 3800260304

18. Liu, J., Liu, X., Lee, E.A.: Modeling distributed hybrid systems in ptolemy ii. Proceedings of the American Control Conference, pp. 4984–4985 (2001)

19. Mitra, D.: Stochastic theory of a fluid model of producers and consumers coupled by a buffer. Adv. Appl. Probab. **20**(3), 646–676 (1988). http://www.jstor.org/ stable/1427040

20. Zanoli, S.M., Barboni, L., Leo, T.: An application of hybrid automata for the mimo model identification of a gasification plant. In: IEEE International Conference on Systems, Man and Cybernetics, 2007, ISIC, pp. 1427–1432, October 2007

A Library of Modeling Components for Adaptive Queuing Networks

Davide Arcelli[1]([✉]), Vittorio Cortellessa[1], and Alberto Leva[2]

[1] Università dell'Aquila, L'Aquila, Italy
davide.arcelli@univaq.it
[2] Politecnico di Milano, Milano, Italy

Abstract. Self-adaptive techniques have been introduced in the last few years to tackle the growing complexity of software/hardware systems, where a significant complexity factor leans on their dynamic nature that is subject to sudden (and sometime unpredictable) changes. Adaptation actions are aimed at satisfying system goals that are often related to non-functional properties such as performance, reliability, etc. In principle, an adaptable software/hardware system can be considered a controllable plant and, in fact, quite promising results have been recently obtained by applying control theory to adaptation problems in this domain.

Goal of this paper is to provide a design support for introducing adaptation mechanisms in Queuing Network models of software/hardware systems. For this goal, we present a consolidated library of modeling components (in Modelica) representing Queuing Network elements with adaptable parameters. Adaptive Queuing Networks (AQN) can be built by properly assembling such elements. Once feedback control loop(s) are plugged into AQNs, it is possible to analyze and control (before the implementation) the system performance under changes due to external disturbances.

We show the construction of an AQN example model by using our library, and we demonstrate the effectiveness of our approach through experimental results provided by the simulation of a controlled AQN.

Keywords: Performance modeling · Control theory · Adaptive systems

1 Introduction

The growing complexity of computing systems is placing increased burden on application developers. This situation is worsened by the dynamic nature of modern systems, which can experience sudden and unpredictable changes (e.g., workload fluctuations and software component failure). It is increasingly up to software/system engineers to manage this complexity and ensure applications operate successfully in dynamic environments [18]. The use of self-adaptive techniques has been proposed to help engineers in managing this burden. Adaptable systems modify their own behavior to maintain goals in response to unpredicted changes. While adaptation of functional aspects (i.e., semantic correctness) often

© Springer International Publishing AG 2016
D. Fiems et al. (Eds.): EPEW 2016, LNCS 9951, pp. 204–219, 2016.
DOI: 10.1007/978-3-319-46433-6_14

requires human intervention, non-functional ones (e.g., performance) represent a challenging opportunity for applying self-adaptive techniques [3]. For example, customers may require continuous assurance of agreed quality levels, which usually map to specific metrics used to trigger adaptations for guaranteeing requirements are met even in the face of unforeseen environmental fluctuations [6]. Such adaptations have been studied for decades in the context of control theory, where control systems have achieved widespread usage in many engineering domains that interact with the physical world [15]. In such domains, a controller measures quantitative feedback from a sensor (e.g., a speedometer) and determines how to tune an actuator (e.g., a fuel intake) to effect the controlled plant behavior (e.g., an engine). One major advantage of using control theory is that such techniques emit analytical guarantees of the system dynamic behavior [21].

In the last few years, control theory has been applied to build adaptable software/hardware systems. Such systems can be in fact considered as controllable *plants* fitting in a basic feedback control loop scheme [15]. The target of a feedback control loop is not necessarily a running system. In fact, control theory is often applied to design system models [13], with the goal of studying (before the implementation) if their dynamic behavior can be controlled/adapted under changes (namely external disturbances), while satisfying non-functional requirements. Here we focus on performance requirements, with the goal of providing support to the design of Queuing Network (QN) models with plugged feedback control loops, for the purposes discussed above. In [4] we presented a first attempt in this direction, where few typical elements of QNs were modeled in Modelica [14], with the purpose of combining them with controllers. QNs assembled on top of these elements have: (i) performance indices, such as response time, which can be observed (as sensed variables), (ii) parameters, such as CPU shares among classes of jobs, that can be modified (as actuators), and (iii) environmental parameters, such as workload and operational profile, which can be modeled as external disturbances whose changes may induce QN adaptations. The first experimental results that we have reported in [4] were promising.

In this paper we consolidate our work by extending the library of Modelica components to represent a large variety of QN elements. We show that such an extended library enables to build a whole class of open QN models that we name here as Adaptive Queuing Networks (AQN). We also show how combining an AQN with controllers allows to investigate the performance of a system under external disturbances and to identify appropriate control laws that have to be implemented in order to guarantee the performance requirements satisfaction through adaptation actions. This paper is organized as follows: in Sect. 2 we discuss related work, Sects. 3 and 4 describe the paper contribution, respectively for the AQN modeling library and for the control modeling, Sect. 5 presents simulation results and Sect. 6 concludes the paper.

2   Related Work

Adaptation is becoming a key concern in software applications [18]. An adaptive application must select, from many configurations, the one that is mostly appropriate to satisfy specific requirements. There are many examples, from hardware to software development. The evaluation of a new microprocessor design requires studying the impact of input data sets and workload composition [12]. Compiler-level advancements have been developed to support adaptive implementations for performance [2] or power [5]. In [10], a study on tuning Fast Fourier Transformations on graphic processing units is presented, whereas Rahman et al. [26] studied the effect of compiler parameters on performance and power consumption for scientific computing.

Besides, control theory [15] is capturing an increasing interest from the software engineering community that looks at self-management as a mean to meet Quality of Service (QoS) requirements despite unpredictable changes of the execution environment [23]. Examples of this trend can be seen in research on control of web servers [22], data centers and clusters management [11], operating systems [8], and across the system stack [17].

The application of control theory in software engineering, however, is still in a very preliminary stage. Developing accurate system models for software is in fact hard, mostly due to the strong mathematical skills needed for dealing with complex non-linear dynamics of real systems [9]. These difficulties usually lead to the design of controllers focused on particular operating regions or conditions and ad-hoc solutions that address a specific computing problem using control theory, but do not generalize. For example, in [16] the specific problem of building a controller for a .NET thread pool is addressed. More in general, the approach presented herein has the peculiarity of not just closing control loops *around* an existing system, or in other terms, adding a control layer *on top* of a fully functional one. Controllers are here part of the system itself, thus in fact enabling elements to provide the required functionality. The interested reader can find in [20, Chapter 1] a discussion on the benefits (and for completeness, the new design challenges) that such an approach brings into the *arena*.

All the related work discussed up to this point aims at controlling running adaptable applications. In this paper, we raise the level of abstraction, in that our contribution concerns *model-based* performance control of adaptive software. In this domain, some effort has been spent to raise adaptation techniques driven by performance (or more general QoS) requirements at the software architecture level [24], where adaptive verification techniques have been also studied [6]. Adaptation approaches for specific architectural paradigms have been introduced, such as Service-Oriented-Architecture [7]. An interesting work has been recently introduced in [27] for automatically extracting adaptive performance models from running applications. However, none of these papers applies control theory to the control of performance models, like we do in this paper.

3   AQN Modeling Library

The first step to define a control methodology for QN models is the provisioning of a suitable mean for their formalization as dynamic systems. To this end, we have extended and consolidated a Modelica library for our purposes, i.e. for enabling the design of AQNs.

Modelica [14] is a modeling and simulation environment widely used by control practitioners to define and study dynamic models of physical phenomena and engineered systems, and to support the design and synthesis of suitable controllers. Our library includes at today several basic QN types, including queues, service centers, routing nodes, workload generators (both deterministic and probabilistic), and it also supports multiple job classes. Instances of these types can be created and connected together to seamlessly draw a QN model. The definition of each type includes peculiar inputs/outputs and state equations. The former determine the interaction with the other components (e.g., input/output rates, control input), while the latter represent a (parametric) dynamic model specified as a system of differential equations that capture the time behavior of the component instance, according to its initial state and the input it receives.

When a QN element is instantiated, the designer needs to set some parameters (e.g., the service rate of a server) and the connection to other components (e.g., reflecting the system control flow). Our library has been conceived to be an instrument that can be extended with other features/types and possibly ported to different contexts. This goal is achieved by exploiting the fact that Modelica is an object-oriented framework with constructs for type hierarchies and inheritance. These mechanisms can be used to extend our library either by adding other QN types (e.g., passive resources) or by extending existing types with additional features (e.g., service centers with special scheduling policies).

We describe in the following all the AQN constructs in the current version of our library that, with respect to the previous one in [4], introduces the following novel aspects: (i) multiple classes of jobs for all elements, (ii) definition of required level of services for each class, (iii) multi-servers for M/M/c models, (iv) ready-to-use predefined stations, (v) more complex split nodes and (vi) corresponding merge nodes.[1]

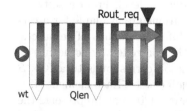

Fig. 1. Station queue.

Station Queue. This element represents a controllable queue of a server, and its state is defined, as usual, by the number of enqueued jobs. The queue capacity is infinite, hence it is possible to push as many jobs as desired. It is also possible to pop an arbitrary number of requests, less or equal than those enqueued, due to the server scheduling mechanism (illustrated later). The graphical representation of a controllable queue, showing its input/output parameters is in Fig. 1. The orange arrow indicates the job flow direction. The main input/output parameters, used for controlling purposes, are explicitly shown. In particular, wt and Qlen represent output "pins" where a controller can be plugged to, in order to observe their current values, i.e., the waiting time per job class and the number of jobs in the queue, respectively. Rout_req, instead, represents an input "knob" of the

---

[1] Note that each parameter described in this section refers to a specific class of jobs, where not differently specified.

queue (i.e., a controlled variable), where again a controller can be plugged to, in order to control the queue length (or the waiting time) with respect to its target value, by actuating on the rate at which jobs are extracted from the queue.

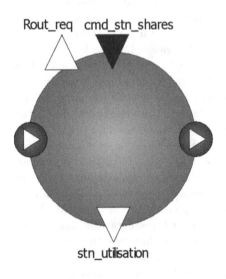

Fig. 2. Station server.

Station Server. This element represents a cluster of S controllable service centers that can process incoming requests at different predefined processing rates for each class of jobs. The graphical representation of a controllable server, showing its input/output parameters, is in Fig. 2, where again jobs flow from left to right. A controllable server has three main parameters: Rout_req and stn_utilisation represent output "pins" which a controller can be plugged to, in order to observe their current values, i.e., its throughput and its utilization, respectively; the input knob cmd_stn_shares can regulate the shares that the server allocates to each class of jobs. We have devised, up today, a single scheduling policy for servers, that is a Generalized Processor Sharing (GPS) one [1]. In particular, we envisage that a (non-uniformly distributed) share of processing is held by each class of jobs, where different classes of jobs may have different resource demands. This is a general modeling approach to represent the case of a service center processing jobs by the same "functional" type (e.g. jobs that represent the same software function/operation) that can be partitioned in classes, where each class provides a different quality level. From a software viewpoint, this is a scenario where different adaptations are available for a certain operation, and each adaptation requires a different amount of resources. The processing shares among classes of jobs may be controlled by actuating on the cmd_stn_shares knob. In particular, such a knob corresponds to a $K \times S$ matrix, namely CMD, where K is the number of job classes. Each element (i, j) of that matrix, with $i = 1..S, j = 1..K$, is in the interval $[0, 1]$ and represents the fraction of the share of the i-th station server in the cluster that is assigned to the j-th job class. To this aim, we have also defined a "pivotal" $K \times S$ matrix, namely MAX_SERV_RATES, where each entry (i, j), with $i = 1..S, j = 1..K$, is a real number greater or equal to zero and represents the *maximum computational effort*, i.e. the maximum rate of the i-th station server in the cluster for the j-th job class. This means that, for example, if the i-th station server has a maximum computational effort of 20 and the fraction of its share for that class is 0.5, then the actual service time of the i-th station server for the j-th job class is 10 jobs per second.

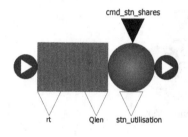

Fig. 3. Station.

Station. This element puts `StationQueue` and `StationServer` together, thus composing a complete service center [19], as illustrated in Fig. 3. A station has three output and one input parameters. The `Qlen` output parameter is inherited from the Station-Queue element, whereas `stn_utilisation` from StationServer. The `rt` output parameter, instead, represents the current residence time for a job in the whole station, as defined in standard QN theory as the sum of waiting time in the queue plus the service time in the server. The input "knob" is inherited from the Station Server.[2]

On the basis of the `Station` element, we have built two types of "ready-to-use" stations, that are shown in Fig. 4. In particular, the one in Fig. 4(a) allows to control the queue length through its `SPql` knob, whereas the residence time is controlled through the `SPrt` knob of the station in Fig. 4(b). Both these `Stations` are equipped with an additional input, named `Auto`, which represents a boolean value that can be manipulated to switch on/off over time the control of these elements, namely "automatic mode".

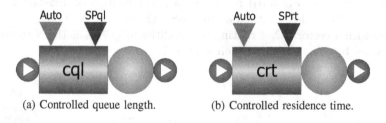

(a) Controlled queue length.   (b) Controlled residence time.

Fig. 4. Predefined `Stations`.

Workload Sources. JobSource element represents an open workload source with two internal parameters corresponding to: (i) the rate at which jobs of each class are generated conforming to a probabilistic distribution (possibly varying over time), and (ii) the service levels requested by each class of jobs (in case one would like to assume that each job class may require a specific service level). Figure 5(a) shows its graphical representation. JobSourceVar, in Fig. 5(b), represents instead a controllable variant of JobSource, where two additional input parameters are introduced, i.e. `rates` and `levels`, representing input knobs for controlling the two above mentioned internal parameters, respectively.

---

[2] Note that in a Station the `Rout_req` input of the StationQueue element has been joined to the homonym output of the StationServer to implement the policy of job extraction from the queue, so they disappear from the figure.

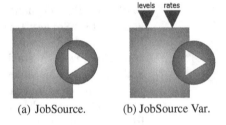

(a) JobSource.   (b) JobSource Var.

Fig. 5. Job sources.

Job Split Nodes. As represented in Fig. 6, these elements are stochastic branchings of jobs, where internally-defined distribution functions regulate the probability that a job is routed along one of the outgoing paths. In particular: JobsSplit2 (Fig. 6(a)) and JobsSplit2Var (Fig. 6(b)) have two outgoing paths and an internal parameter, common to all job classes, corresponding to the probability that a job is routed along the first path (i.e., the one on the top). In addition, JobsSplit2Var exposes the p input parameter for externally setting such probability. JobsSplit2_probPerClass (Fig. 6(c)) is similar to JobsSplit2, but it allows to specify a vector of probabilities (namely ppc), i.e. one for each job class. Finally, JobsSplitN (Fig. 6(d)) represents a generalization of JobsSplit2, i.e. a N-way job splitting; hence, it allows to specify the number N of outgoing paths, together with a vector of N-1 routing probabilities $p_1..p_{N-1}$, as the N-th one can be obtained by complementing their sum to 1.[3]

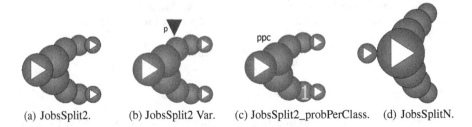

(a) JobsSplit2.   (b) JobsSplit2 Var.   (c) JobsSplit2_probPerClass.   (d) JobsSplitN.

Fig. 6. Job splitting.

Note that split node probabilities can be associated to either software characteristics or hardware ones. The former case is introduced in the example QN of Sect. 5. The latter case can be used for routing control in the platform, and in particular for load balancing purposes.

---

[3] For sake of simplicity, only p has been defined as a knob in the current implementation of our library. The library is open to make all other probabilities as knobs.

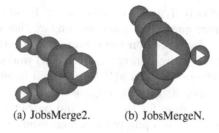

(a) JobsMerge2.   (b) JobsMergeN.

Fig. 7. Job merges.

Jobs Merge Nodes. As graphically represented in Fig. 7, these elements allow to merge previously split jobs. In particular, JobsMerge2 (Fig. 7(a)) can be coupled to any split among JobsSplit2, JobsSplit2Var, and JobsSplit2_probPerClass, whereas JobsMergeN (Fig. 7(b)) couples to JobsSplitN.

4   Assembling AQN and Controllers

4.1   AQN Assembly

There are several ways to model a QN as a dynamic system suitable for the application of control-theoretical management techniques. The one we refer to herein is the combination of two entities. The first one is a set of dynamic systems representing each a processing node, irrespectively of its position in the network. The second one is a probability transition matrix (i.e., a non-dynamic object) that dictates the network topology and the job routing.

The approach just sketched can be applied in both the continuous and the discrete time. In this paper we choose the first case, which implicitly corresponds to assume that job rates are high enough that representing in the model the history of each individual job would be impractical and computationally not advisable. Incidentally, the use of continuous-time models allows to exploit variable-step solvers [25], which can significantly speed up simulation.

The dynamic systems corresponding to nodes, in turn, are the compound of a queue and a server that takes jobs from the queue. The dynamic character comes form the queue, where jobs can accumulate, while the operation of the server is memoryless (i.e., the time to process a job does not depend on the past history of the server). If the queue – for which we assume infinite capacity – does not get emptied, the node model is linear and time-invariant. The transition matrix can be constant, thereby not destroying linearity and time invariance, or some probabilities may be time-varying, in which case the latter property is lost.

If there is more than one job class, all the above is repeated for each of them. In particular, a node has a server for each class it manages, and the probabilities in the transition matrix are expressed for each class—equivalently, the said matrix has a third dimension, and its size in that dimension equals the number of job classes.

As anticipated, the only dynamic element in the library, except controllers, is the model of the server queue. For space reasons we thus concentrate on that model, spending just a few words on the server, and devoting a more comprehensive description of the entire library to specialized works.

We assume the queue occupation $n(t)$ to be a real number, given again the hypothesis of "large" amounts of jobs (e.g., n could be measured in *kilo*jobs). We make the same assumption for the input and output rates $r_i(t)$ and $r_o(t)$. In addition we consider $r_i(t)$ as a purely exogenous input, while we assume that for $r_o(t)$ a required value $r_{o,req}(t)$ comes from the job server. The queue thus emits jobs at (instantaneous) rate $r_{o,req}(t)$ if there are some, otherwise the actual output rate $r_o(t)$ is zero. This immediately reflects into the general differential-algebraic model $\frac{dn(t)}{dt} = r_i(t) - r_o(t)$. Such model can be detailed as follows:

$$\frac{dn(t)}{dt} = \begin{cases} 0 & n(t) \le 0 \text{ and } r_i(t) - r_{o,req}(t) < 0 \\ r_i(t) - r_{o,req}(t) & otherwise \end{cases} \tag{1}$$

As for the server, the model just consists of two computations:

- determine the actual CPU shares by taking the commanded ones, either constant or dynamically computed by a controller, and managing possible overutilizations (in this paper we simply scale the shares linearly if their sum exceeds the unity, but there are plenty of alternatives);
- determine the required output rate for each job class queue by dividing the corresponding share by the (assumed) processing time for that job class.

4.2   Control Modeling

In this paper we propose a decentralized control strategy, i.e., each node has a local controller and these do not communicate with one another. The purpose of this controller is to regulate the length of the node's queue, job class by job class, by acting on the shares of the node CPU devoted to the servers for the classes. In the treatise we try to use as few control-theoretical concepts as possible, but nonetheless a minimum background on the matter is advisable. The reader needing such information in a compact form can refer, e.g. to [20, Chapter 2 and 4]

In the Laplace transform domain, and limiting the notation to one job class for lightness, the linear behavior of a queue is represented by the transfer function

$$N(s) = P(s)\big(R_i(s) - R_o(s)\big), \quad P(s) = \frac{1}{s}, \tag{2}$$

where $N(s)$ is the transform of the occupation and $R_i(s)$, $R_o(s)$ respectively those of the input and the output job rate. The latter rate is the control variable, as it can be altered by acting on the server CPU share, while the former rate is a disturbance for the local controller $C(s)$. Selecting a PI structure for that controller, we have

$$R_o(s) = C(s)\big(N^\circ(s) - N(s)\big), \quad C(s) = -K\left(1 + \frac{1}{sT_i}\right), \tag{3}$$

where $N°(s)$ is the transform of the desired (set point) occupation, $T_i > 0$ and $K > 0$, both for the Bode criterion and – more intuitively – because to increase the queue occupation one has to decrease the output rate. The open-loop transfer function is thus

$$L(s) = K \frac{1 + sTi}{s^2 T_i}.$$ (4)

Accepting to evaluate the cutoff frequency ω_c and the phase margin φ_m on the asymptotic Bode diagrams, we obtain

$$\omega_c = \begin{cases} \sqrt{K/Ti} & K \leq 1/Ti \\ K & K > 1/Ti \end{cases} \quad \varphi_m = \begin{cases} \arctan \sqrt{KTi} & K \leq 1/Ti \\ \arctan(KTi) & K > 1/Ti \end{cases}$$ (5)

To select K and T_i, recall that the closed-loop settling time t_s (which is the time to recover the occupation set point after a step-like disturbance) approximately equals $5/\omega_c$, while the phase margin is a measure of the loop stability degree and robustness. In this case robustness is not an issue as the model of the controlled system (2) is uncertainty-free. However it is advisable to avoid oscillatory responses, hence to have a "high" φ_m. Since the two cases in (5) correspond respectively to the φ_m ranges $(0°, 45°]$ and $(45°, 90°)$, we select the second one. Given desired values $\bar{t}_s$ and $\overline{\varphi}_m \in (45°, 90°)$ for t_s and φ_m, tuning is performed by setting

$$K = \frac{5}{t_s}, \quad T_i = \frac{t_s}{5} \tan \overline{\varphi}_m.$$ (6)

The local controller can either be used to just maintain the queue occupation below a certain level, or can have its set point dynamically calculated as a desired waiting time multiplied by the measured inlet rate. In this case, the effect is to adapt the server processing speed so as to guarantee that the waiting time does not exceed the set point (it can of course be lower if there are no waiting jobs, but this is managed naturally by the PI saturation and anti-windup mechanism).

5 Evaluation

In this section we show the presented Modelica library at work by representing and simulating a network subjected to disturbances.[4] In the example we also briefly illustrate how a quite basic control scheme can improve the behavior of the network, dynamically allotting computational resources so as to maintain a prescribed operation quality in the face of the mentioned disturbances.

The Modelica diagram for the network is shown in Fig. 8, and describes a simple online shop. There are three job classes, namely MakePurchase, Browse-Catalog, and Register. All enter the network through the Server node, and then

---

[4] A resource can be downloaded at http://www.di.univaq.it/davide.arcelli/resources/ EPEW2016.zip including the current version of our library and the usage example in this paper.

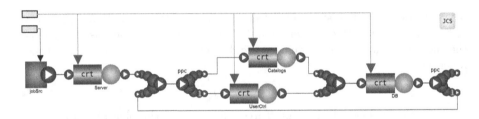

Fig. 8. Simulation example – Modelica diagram for the considered network.

are split: MakePurchase and BrowseCatalog jobs go to the Catalogs node, while Register jobs are served by the UsertCtrl one. All jobs then go to a database (DB) node. After that, jobs of Register, MakePurchase, and BrowseCatalog, are recycled to merge with the output of the Server node, with probabilities equal to 0.1, 0.3, and 0.5, respectively.

The network is subjected to time-varying inputs, as shown in Fig. 9. In this figure and the following ones the job classes are numbered, 1, 2, and 3 corresponding to MakePurchase, BrowseCatalog and Register, respectively. All the input rates are composed of a constant baseline value, to which a disturbance made of a double (up/down) step plus a sine-like variation, is superimposed three times in the simulated experiment duration. For sake of simplicity, we do not assume each job class may require a specific service level. Table 1 summarizes the cost in terms of CPU time for processing one job of each class on the four nodes in the network (recall that not all nodes serve all job classes).

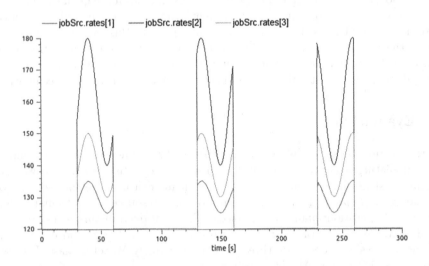

Fig. 9. Simulation example – input rates.

Table 1. Computational cost of one job of the classes on the network nodes.

	MakePurchase (s)	BrowseCatalog (s)	Register (s)
Server	0.015	0.015	0.015
UserCtrl	n/a	n/a	0.065
Catalogs	0.5	0.2	n/a
DB	0.3	0.9	0.3

The network is operated in automatic mode for $t < 100$ and $t \geq 200$, while for $100 \leq t < 200$ it is disabled. In automatic mode, each station is governed by a local controller to guarantee a job residence time (i.e., waiting plus processing time) not exceeding 100ms; to do that, the controller can act on the shares of the station CPU allotted to each job server aboard it. When automatic mode is disabled, conversely, each job class server is allotted a fixed station CPU share. The said shares were computed statically for the baseline input rates, plus some over-provisioning for the safe side, with the additional constraint of not exceeding 60 % of the entire station CPU availability.

Figure 10 shows the residence time for each class of job on each node in the network. When in automatic mode, local controllers keep the required residence times. The match to the required value is very good, if not for some effects of the step-like components of the input rate disturbances—the harshest possible *stimulus* in this respect, anyway. All in all, if the required residence times were part of a service level agreement, violations would be practically negligible.

Things are quite different when automatic mode is disabled. The statically allotted CPU shares, given the necessary over-provisioning, cause the residence time with the baseline input rates to be lower than required, which is plain obvious. However, since a reasonable over-provisioning was applied not to unduly steal resources for possible other applications using the nodes, as soon as the disturbance becomes too significant – albeit tractable, as shown by the operation in automatic mode – the requirement on residence times cannot be fulfilled anymore. In the absence of feedback, the network thus drifts away from the goal, as shown in Fig. 10(d) for the DB node. The network moves back toward the goal only as the disturbance goes away—or, more in general, comes back within a range that the *a priori* over-provisioned resources can manage. Of course one may think of adapting the over-provisioning amount from time to time based on some analysis of the past behavior and/or some forecast, but as can be seen, feedback control does the job with less complexity than any mechanism of the type just envisaged.

Carrying, Fig. 11 shows the CPU shares allotted to each job class server on each node in the network. The first thing to notice is that with a thoroughly performed tuning – which we do not discuss here as doing so would require space and stray from the scope of this paper – the action of the proposed control, that it is worth recalling to be completely reactive, is fast enough to contrast disturbances very effectively. In other words, considering reactive policies to be

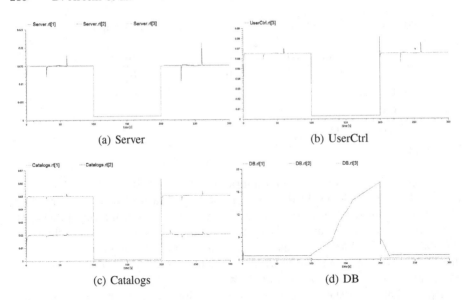

Fig. 10. Simulation example – residence times for each job class on the four nodes.

"slow" irrespective of how they are set up and tuned, is a conceptual error. If one adds that reactive policies are the only ones usable in the absence of *reliable* forecasts, and considers how difficult it can be to ensure the reliability of a forecast, then many designs can benefit. Moreover, referring again to Fig. 11, the

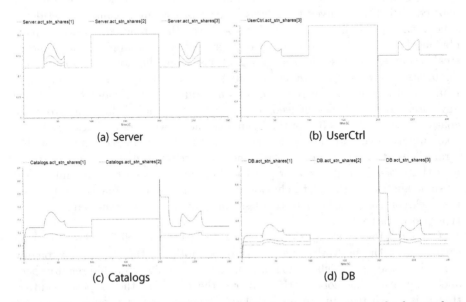

Fig. 11. Simulation example – CPU shares for each job class server on the four nodes.

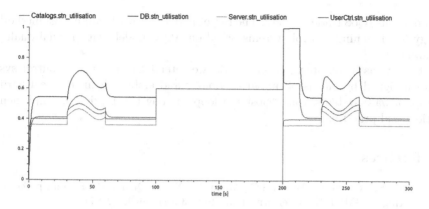

Fig. 12. Simulation example – total CPU utilizations of the four nodes.

applied local control is inherently capable of moving resources from one server to another when needed, i.e., to not unduly over-provision some server when some other is starving.

To complete the presentation and the statements just made, Fig. 12 reports the CPU utilization for the network nodes. It is worth to be noticed that for most of the time, including when the disturbance is acting, the CPU utilizations under automatic mode are lower than those with static provisioning, and only very sparingly go above the 60 % assumed bound. The advantages of this should be apparent in terms of both management simplicity and economy, as no complicated evaluations are necessary to find the best (minimum) over-provisioning, and resources are not allotted unless really needed.

6   Conclusion and Future Work

In this paper, we have presented a library of Modelica components representing QN elements with adaptable parameters. The design of a simple controller have allowed to show an usage example of such library, where an AQN has been built with embedded controllers for satisfying performance goals under disturbances.

We like to remark that our library has been conceived to be open and extensible, so that performance experts and control engineers can define new modeling components both for the AQN and the control modeling sides. For example, in this paper we applied the control mechanism to single QN stations, and controllers do not communicate. A significant short-term extension of our library would be the introduction of hierarchic control for enabling centralized strategies aimed at global performance requirements fulfillment. Another extension that we intend to introduce in the near future would be to support also finite queue capacities, in addition to infinite ones that we have addressed so far.

In the mid-term future, we will relay on our library for tackling many more control problems of AQNs, such as: load balancing and allocation of virtual machines, software, and hardware resources, admission control, temperature/QoS trade-offs, etc. The target here is a broadly applicable methodology for designing control systems based on AQN models with formal quality guarantees.

Finally, as a long-term future work, we intend to implement control systems designed by means of our methodology and embed them into real hardware/software systems, thus "closing a loop" among theory, design, implementation, and practice.

References

1. Aalto, S., Ayesta, U., Borst, S.C., Misra, V., Núñez-Queija, R.: Beyond processor sharing. SIGMETRICS Perform. Eval. Rev. **34**(4), 36–43 (2007)
2. Ansel, J., Chan, C., Wong, Y.L., Olszewski, M., Zhao, Q., Edelman, A., Amarasinghe, S.: PetaBricks: a language and compiler for algorithmic choice. In: ACM PLDI (2009)
3. Arcelli, D., Cortellessa, V.: Challenges in applying control theory to software performance engineering for adaptive systems. In: ICPE, pp. 35–40 (2016)
4. Arcelli, D., Cortellessa, V., Filieri, A., Leva, A.: Control theory for model-based performance-driven software adaptation. In: QoSA 2015, pp. 11–20 (2015)
5. Baek, W., Chilimbi, T.: Green: a framework for supporting energy-conscious programming using controlled approximation. In: ACM PLDI (2010)
6. Calinescu, R., Ghezzi, C., Kwiatkowska, M., Mirandola, R.: Self-adaptive software needs quantitative verification at runtime. Commun. ACM **55**(9), 69–77 (2012)
7. Cardellini, V., Casalicchio, E., Grassi, V., Iannucci, S., Presti, F.L., Mirandola, R.: MOSES: a framework for QoS driven runtime adaptation of service-oriented systems. IEEE Trans. Softw. Eng. **38**(5), 1138–1159 (2012)
8. Cascaval, C., Duesterwald, E., Sweeney, P.F., Wisniewski, R.W.: Performance and environment monitoring for continuous program optimization. IBM J. Res. Dev. **50**(2/3), 239–248 (2006)
9. Dorf, R., Bishop, R.: Modern Control Systems. Prenntice Hall, Upper Saddle River (2008)
10. Dotsenko, Y., Baghsorkhi, S.S., Lloyd, B., Govindaraju, N.K.: Auto-tuning of fast fourier transform on graphics processors. In: PPoPP, pp. 257–266 (2011)
11. Dutreilh, X., Moreau, A., Malenfant, J., Rivierre, N., Truck, I.: From data center resource allocation to control theory and back. In: CLOUD, pp. 410–417 (2010)
12. Eeckhout, L., Vandierendonck, H., Bosschere, K.D.: Quantifying the impact of input data sets on program behavior and its applications. J. Instr.-Level Parallelism **5**(1), 1–33 (2003)
13. Filieri, A., Ghezzi, C., Leva, A., Maggio, M.: Self-adaptive software meets control theory: a preliminary approach supporting reliability requirements. In: ASE, pp. 283–292 (2011)
14. Fritzson, P., Engelson, V.: Modelica: A unified object-oriented language for system modeling and simulation. In: Jul, E. (ed.) ECOO. LNCS, vol. 1445, pp. 67–90. Springer, Heidelberg (1998)
15. Hellerstein, J.L.: Self-managing systems: a control theory foundation. In: ECBS, pp. 708–708 (2005)
16. Hellerstein, J.L., Morrison, V., Eilebrecht, E.: Applying control theory in the real world: experience with building a controller for the.net thread pool. SIGMETRICS Perform. Eval. Rev. **37**(3), 38–42 (2010)

17. Hoffmann, H., Holt, J., Kurian, G., Lau, E., Maggio, M., Miller, J., Neuman, S., Sinangil, M., Sinangil, Y., Agarwal, A., Chandrakasan, A., Devadas, S.: Self-aware computing in the angstrom processor. In: DAC (2012)
18. Kramer, J., Magee, J.: Self-managed systems: an architectural challenge. In: FOSE, pp. 259–268 (2007)
19. Lazowska, E., Kahorjan, J., Graham, G.S., Sevcik, K.C., Performance, Q.S.: Computer System Analysis Using Queueing Network Models. Prentice-Hall Inc., Upper Saddle River (1984)
20. Leva, A., Maggio, M., Papadopoulos, A.V., Terraneo, F.: Control-based Operating System Design. Institution of Engineering and Technology, London (2013)
21. Levine, W.: The Control Handbook. CRC Press, Boca Raton (2005)
22. Lu, C., Lu, Y., Abdelzaher, T., Stankovic, J., Son, S.: Feedback control architecture, design methodology for service delay guarantees in web servers. IEEE Trans. Parallel Distrib. Syst. 17(9), 1014–1027 (2006)
23. Patikirikorala, T., Colman, A., Han, J., Wang, L.: A systematic survey on the design of self-adaptive software systems using control engineering approaches. In: SEAMS, pp. 33–42 (2012)
24. Perez-Palacin, D., Mirandola, R., Merseguer, J.: On the relationships between QoS and software adaptability at the architectural level. SoSyM J. 87, 1–17 (2014)
25. Petzold, L.R., et al.: A description of DASSL: a differential/algebraic system solver. In: Proceedings of IMACS World Congress, pp. 430–432 (1982)
26. Rahman, S.F., Guo, J., Yi, Q.: Automated empirical tuning of scientific codes for performance and power consumption. In: HiPEAC, pp. 107–116 (2011)
27. Zheng, T., Litoiu, M., Woodside, C.M.: Integrated estimation and tracking of performance model parameters with autoregressive trends. In: Kounev, S., Cortellessa, V., Mirandola, R., Lilja, D.J. (eds.) ICPE, pp. 157–166. ACM, New York (2011)

Author Index

Printed in the United States
By Bookmasters